DAYLIGHT DESIGN OF BUILDINGS

DAYLIGHT DESIGN OF BUILDINGS

Nick Baker and Koen Steemers

Published by James & James (Science Publishers) Ltd
35–37 William Road, London, NW1 3ER, UK

A catalogue record for this book is available from the British Library.

ISBN 1 873936 88 5

Daylight Design of Buildings documents a major outcome of a comprehensive
European research program, Daylight Europe JOU2-CT94-0282, 119-97, a project of the European JOULE III Research Program. The project was funded by
the European Commission DG XII for Science, Research and Development, Rue
de la Loi 200, B-1049 Brussels, Belgium, and individual member states participating in the project.

Printed in Hong Kong by H & Y Printing Ltd

Cover photograph: Internal view of Trinity Library, Cambridge. Courtesy of
Øyvind Aschehoug.

Contents

Acknowledgements

The European Commission has funded research and development into the improved energy efficiency and environmental performance in buildings for more than two decades. In particular the EU Daylight Europe project (Joule II CT 94-0282) provided the inspiration and the majority of the source material for this book. Firstly we would like to acknowledge Dr Georges Deschamps, our project supervisor from the European Commission DGXII, and secondly Poul E Kristensen who, as one of the participants, was the scientific co-ordinator of the project.

The project involved 22 participating groups from 13 countries. These groups have contributed directly with material from the 60 case studies, published in detail separately under the title *Daylight Performance of Buildings*. Many participants also contributed to the shaping and development of this document during discussions held at project meetings, and in written comments on early draft material.

More specific contributions as authors have been made as follows:

Raphaël Compagnon	Chapter 9 and 10.1
David Crowther	Chapter 10.2
Paul Littlefair	Chapter 8
Øyvind Aschehoug	Chapter 2.1, 2.3 and Chapter 11.1.
Laurent Michel	Chapter 11.4
Katerina Parpairi	Chapter 13

Marc Fontoynont and André de Herde also contributed material to Chapters 6 and 4 respectively. Simulation results have been provided by Joe Clarke.

In their role on the Design Guidelines steering committee we also acknowledge the valuable comments and suggestions of Alexandros Tombazis, Helena Coch and Bernard Paule.

Finally, vital contributions have been made by Katerina Parpairi, coordinating the production of this draft document as well as making numerous direct contributions to it, and Mike Baker for the graphic design and drawing the majority of the diagrams.

Introduction

"Daylight is a gift of Nature. As civilised man learns to use artificial light sources which free him from total dependence on daylight, he also learns to appreciate the value of daylight and become aware of its special advantages."

Hopkinson wrote these words in 1966, in the opening chapter of what is now regarded as the 'scientific Bible' of daylighting design[1]. In spite of the sentiments expressed having resonance in the minds of the architect, as well as the engineer and scientist, Hopkinson's book is little known by the design profession and has had little direct influence on the actual design of buildings. Why is this?

Many architects would regard daylighting design as a matter of common sense and experience. It is primarily seen as a problem of facade composition, and the architect uses precedent, exemplars and stylistic rules for support. However, the widespread acceptance of the use of artificial lighting during daytime acts as a fail-safe, often masking the lack of success of the daylighting design. Only with the growing interest in energy saving by the use of daylight, and an increased awareness of visual comfort, is the success of daylighting design being viewed more critically.

A critical and objective view such as this has been applied in 60 European building case studies, described in the companion volume *Daylight Performance of Buildings*[2]. Here in *Daylight Design of Buildings* we provide a more synthetic approach, developing the scientific principles that lie behind successful daylighting design. However, to keep this within a real context, we illustrate the principles with examples drawn almost entirely from the case study buildings. Furthermore, the scientific principles are put forward not for their own intrinsic interest (although this may exist for many readers) but only in so far as they assist in understanding, and ultimately improve the quality of the design solution.

The traditional approach to design can be likened to an evolutionary process – small changes, often made for arbitrary reasons, are reviewed and judged, and the more successful retained. Where rates of change are relatively slow this can be satisfactory. However, due to the long timescale of building design, production and use, the feedback loop is slow. When, as we have seen in the past 50 years, both functional constraints and technical opportunities change rapidly, we need the support of analytical methods. For example we have no 'traditional response' to the daylighting requirements for a room full of VDUs; similarly few designers could claim that they have intuitive knowledge of how to design a lightshelf or specify a lighting control system.

The objective of this book is to provide the necessary tools to assist the designer to provide well-daylit interiors. At the same time it seeks to show that good daylight design is not a restriction on architectural expression – on the contrary, it is more likely to act as an inspiration and foundation for good architecture. In this respect, Part 1 provides a historical overview, placing developments in daylighting design on a timescale, and against the background of climate, building use and cultural values. It demonstrates the huge impact that the provision of daylight within a building has had upon the architectural expression and meaning, at all levels.

Part 2 is the main technical section, following a logical structure that moves from the geographical location, the site and form of the building, through room design to the detailed design of the windows including advanced daylighting elements. At the end of each chapter there is a summary that contains the essential points. The sequential structure does not however imply that the reader has to work rigorously through the sequence. By use of cross-referencing, the glossary, and the summary pages, it should be possible to join the text at any point.

Part 3 contains much of the data and information referred to in the earlier text. The authors have attempted to include sufficient data to carry out the more basic daylight design calculations, e.g. daylight factor and daylight sufficiency. This section also contains information on daylighting design tools, both manual and computer-based. However, quite a lot of useful data (tables and graphs) will be found in the main text of Part 2 illustrating the relevant topics.

Finally, the authors are aware of the difficulty in achieving comprehensiveness. The book is the result of inputs from many individuals and institutions, spread over a lapsed time of several years. However, in spite of limitations we hope that the book contributes to re-kindling an interest in daylighting design, and provides sufficient technical information and analysis, to bring the designs to a successful realisation.

References

1   R G Hopkinson, P Petherbridge and J Longmore, *Daylighting* (London: Heinemann, 1966).

2. M Fontoynout (ed), *Daylight Performance of Buildings* (London: James & James (Science Publishers) Ltd / EC DGXII,1999).

Referencing convention: Bibliographic references appear as superscript numbers. Cross-referencing to other parts of this book appears in round brackets viz (2.3).

Part 1

The role of light in architecture

1 The role of light in architecture

Bibliothèque
National, Paris, by
Labrouste, 1859–68.
(James Austin,
Cambridge)[38]

The role of light in architecture

"... Architecture is the masterly, correct and magnificent play of volumes brought together in light..."
"The history of architecture is the history of the struggle for light."

Le Corbusier[1]

The roles of daylight in architecture are considered from two basic perspectives in this chapter: that of art and science, emotion and quantity, or heavenly and earthly. Historically these roles have been intertwined, but since the age of Enlightenment they have begun to separate and become distinct. In that separation we have lost a holistic understanding of the role of light, but gained knowledge of the science of light. It is through the application of this knowledge to scrutinise the qualities of architectural masterpieces that we can hope to acquire the artistry of lighting design. The purpose of this endeavour is to fulfil both the personal emotional needs – of well-being, comfort and health – and the practical communal needs – of energy conservation – to underpin an environmentally sustainable future.

The understanding and manipulation of light goes to the heart of the architectural enterprise. Vision is the primary sense through which we experience architecture, and light is the medium that reveals space, form, texture and colour to our eyes. "More and more, so it seems to me, light is the beautifier of the building" (Frank Lloyd Wright). More than that, light can be manipulated through design to evoke an emotional response – to heighten sensibilities. Thus architecture and light are intimately bonded.

However, light is not only related to the visual experience of form and space, but is strongly connected to thermal qualities. Light is energy, and whether diffuse or direct, will change to heat when it falls on a surface. This is most noticeable when the light source is intense, as with direct sunlight. The implications of this for the sensorial environmental characteristics of a space are numerous and have a direct impact on air and surface temperatures, as well as indirectly on thermal comfort and air movement. The characteristics of light, heat, air movement and comfort are the key factors in determining a building's energy consumption, and if manipulated and controlled correctly will minimise reliance on artificial systems.

Thus light in architecture is not of singular concern that can be isolated from other design concerns, but relates to a rich integrated web of interdependent aesthetic and functional criteria.

It is through the critical appreciation of precedents, and the understanding of the science of light, that the art of lighting in architecture can be better instilled in the design process. An investigation of the roles of light through history reveals both the power and beauty of light in architecture. The manipulation of daylight through building designs of our architectural heritage shows the ingenuity of integration and the resolution of the perceived struggle between aesthetic purpose and technical understanding of light.

It is possible to trace the roots of the meaning of light back to early civilisation and before. Through the evolution of man and of society, light is associated with safety, warmth and community. Daylight gave man expansive views over his territory. At night, the flame defined a social focus, allowed man to see in the dark where dangers lurked, and to be protected from the cold. Buckminster Fuller eloquently links sun and fire: "Fire is the sun unwinding from the tree's log".[2] The strong relationship of light with life is likely to be the reason for its spiritual associations today. This is revealed most powerfully through the religious and ceremonial architecture of past civilisations. However, it is not only in the exclusive context of 'high' architecture that an artful skill in daylight design can be witnessed, but also in the more widespread and often functionally led vernacular and 'primitive' designs. Put simply, high architecture is concerned with the lighting of the architecture, whereas vernacular buildings are predominantly concerned with bringing daylight in to enable activities to take place – the lighting of a task. Notwithstanding this polarity of purpose, it is frequently the vernacular that produces the spectacular, particularly in terms of the embodied wisdom that is represented.

1.1 Spectacular vernacular

"Thousands of years of accumulated expertise has led to the development of economic building methods ... climatization ... and an arrangement of living and working spaces in consonance with their social requirements. This has been accomplished within the context of an architecture that has reached a very high degree of artistic expression."

Hassan Fathy[3]

Figure 1.1 The fortified cathedral at Albi, France (begun 1282) seen against the vernacular buildings of the town. (Wim Swaan, London)[4]

Figure 1.2 Design of traditional structures responding to the needs of climate and culture. Unusual roofscapes, typical of the Sind district in Pakistan, which channel the wind into the building. (Courtesy Atlantis Verlag, Zurich)[5]

The vernacular tradition has much to teach the modern designer, particularly in response to climatic parameters, notably sunlight with all its visual, thermal, and energy implications. It is also the background against which more monumental architecture is seen and understood (Figure 1.1). Although 'high' architecture demonstrates the level of sophistication, power and standing of the elite, traditional design is more closely related to the views and needs of all people, and thus an unselfconscious expression of the society and its culture (Figure 1.2).

1.1.1 Shade and air

In extreme climates, particularly where the sun's light and heat are too intense, the vernacular and even primitive builders have turned their hand and initiative to the architectural expressions of solar control.

In warm humid climates primitive dwellings typically consist of a raised open stick construction with a wide over-sailing roof to keep the sun off (Figure 1.3). The

Figure 1.3 A tree house in the village of Buyay, located on Mount Clarence in New Guinea, making the most of the faintest forest breeze.[6]

Figure 1.4 The simple timber or bamboo envelope, typical of traditional architecture in warm humid climates, reduces glare, diffuse solar gains and provides privacy while allowing good cross-ventilation. A wide overhang provides the primary shading from sunlight (Vernacular Buildings Museum, Hainan Province, South China).

open lattice bamboo walls reduce the intense glare of reflected light, limiting visual discomfort, whilst allowing the cooling breeze through to the occupants (Figure 1.4).

A similar lattice, but made of highly crafted timber, is often seen within the openings in hot arid climates. Called a *mashrabiya*, the environmental role of the lattice

Figure 1.5 An outside view of a *mashrabiya* at the As-Suhaymi house in Cairo, showing the privacy of the interior.[7]

Figure 1.6 View from a *mashrabiya*. The camera is positioned inside the room but focused on the building across the courtyard, showing the views allowed of more public spaces.[8]

is to soften the light (particularly when made of turned wood – round in section) and allow air flow through whilst maintaining a combination of privacy and a view out. Privacy and view out are the result of the relative intensities of light on either side of the screen, where from the lighter side you see the screen but are unable to see detail through it to the darker interior, thus providing privacy (Figure 1.5). Conversely, from the darker side the screen allows views through to the lighter and more public spaces (Figure 1.6). Furthermore, the size of the interstices between the wood pieces is such as to minimise direct high angle sun penetration, but to transmit some diffuse reflected light perpendicular to the *mashrabiya*. At eye level the spacing is reduced to limit glare, but higher up the spacing is increased to allow daylight deeper into the space (Figure 1.7). Apart from such screened openings, other less important windows are kept small and set with deep reveals in the thick masonry walls, helping to minimise glare and sunlight penetration.

Figure 1.7 The interior view of the *mashrabiya* of Figure 1.5. It covers the entire facade of a room in the house and increases ventilation while reducing glare.[9]

1.1.2 Urban vernacular

The vernacular urban architecture of predominantly hot arid climates, such as Egypt, is characterised by dense courtyard forms, constructed of thick masonry walls with a predominance of small window openings. The proximity of the dwellings to each other ensured that the narrow streets remained sheltered from the intense direct

Figure 1.8 Plan of part of the village of Baris, Al-Kharga Oasis, Egypt, with its multiple shaded and exposed courtyards.[10]

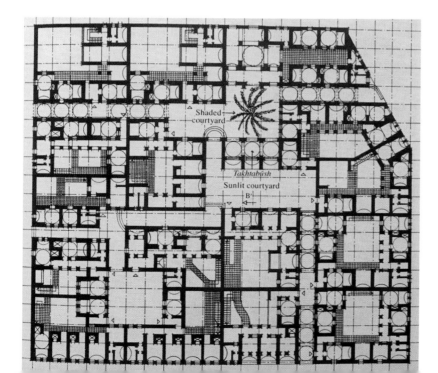

sunlight, and that buildings would provide mutual shade (Figure 1.8).

The private organisation of the dwelling around the focus of a deep shady courtyard, often planted and with the cooling effects of a small fountain, allowed a range of social and working activities to take place outside – including sleeping – in privacy and security during hot summer months. During the daytime the diffuse light from the courtyard can be allowed into the interior via large *mashrabiya*-filled openings with minimum glare problems.

1.1.3 Sun and shade

As we move to southern Europe, away from the extreme climatic conditions of northern Africa, the climate is milder and has distinct seasonal variations, with warm summers and cooler winters.

The sun's energy needs to be selectively controlled through design to ensure shade in the summer and solar warmth in the winter. This was particularly important where alternative fuel sources, notably timber, were either becoming scarce or were needed for construction (of buildings and more importantly of ships). In Greece and Italy, the value of the sun became well understood and its manipulation through building design became highly refined.

In ancient Greece, the naturalist Theophrastus noted that "the sun provides the life-sustaining heat in animals and plants", and exposure to sun was considered good for health. The medical authority Oribasius claimed that north-facing areas were healthier, compared with the south, because they "do not receive much sun and when they do, the light falls obliquely without much vitality".[11]

The Greeks' use of sundials made them aware of solar geometry and how it could be related to building design. The orthogonally planned cities of Olynthus and Priene consisted of house plans organised around south-facing courts and porticoes. Socrates described the concept:

"In houses that look toward the south, the sun penetrates the portico in winter, while in summer the path of the sun is right over our heads and above the roof so that there is shade".[12]

Thus there is a subtle but significant variation from the Egyptian house with courtyard and *mashrabiya* keeping direct sunlight out at all times, to the Greek dwelling with 'atrium' and portico, which selectively allows winter sun (Figure 1.9). The critical first steps in passive solar design were thus established.

1.1.4   Orientation

Vitruvius, the pre-eminent Roman architect of the first century BC, enunciated the need for a relationship between the design of dwellings and the sun's path:

"One type of house seems appropriate for Egypt, another for Spain ... one still different for Rome... This is because one part of the earth is directly under the sun's course, another is far away from it, while another lies midway between the two ... It is obvious that designs for homes ought to conform to the diversities of climate".[14]

Dwellings for the wealthy were planned in accordance with the solar principles established by the Greeks and described in detail by Vitruvius. Even the orientation of specific rooms was outlined by Vitruvius, who for example recommended that winter dining rooms and baths looked towards the winter sunset "because the setting sun, facing them in all its splendour but with abated heat, lends a gentler warmth to that quarter in the evening".[15] Transparent window coverings became available, initially in the form of thin stone (mica or selenite) and by the first century AD in the form of clear glass. Seneca, the philosopher, records this development in AD 65:

"Certain inventions have come about within our own memory – the use of window panes which admit light through transparent material, for example".[16]

The use of glass enabled openings to be larger without letting the warm interior air escape. This, together with the structural innovations of the Romans, allowed large amounts of daylight into their buildings. This became an increasing asset as the Roman Empire reached northwards.

1.1.5   Temperate Europe

In the more temperate climates of Europe it is noteworthy that an equivalent amount of effort is invested in the design and construction of the key moderators of light. The need for shelter from high summer sun in southern Europe is provided by the primitive yet effective south-facing loggia of the Greek dwelling, or the urban arcades of Spanish towns. In northern Europe, the

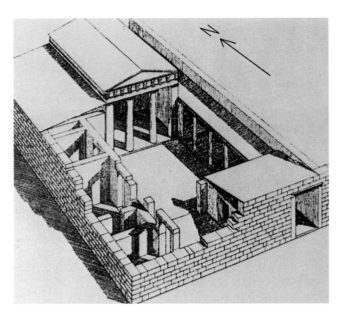

Figure 1.9
Reconstruction of a classical Greek home, from excavations at the city of Priene. The rooms behind the portico faced south onto the courtyard.[13]

aim has been to allow the winter sun in whilst keeping the cold out through window designs and shutters. It is typically these elements of environmental control, as well as the fireplace and its chimney, that receive design and aesthetic attention in vernacular buildings. Even the most humble dwelling required a careful consideration of lintel design, typically made of superior quality stone brought from further afield, to carry the more random quality local materials of the wall above. Windows would become increasingly elaborate with increasing wealth. The simple rectangular opening consisted of vertical timber or stone mullions and timber shutters – adequate for basic ventilation control. Later, the use of oiled paper, and subsequently glass panes by the middle of the 17th century, allowed light in but not cool air through the tall openings with more elaborately carved surrounds.

Most of life took place outdoors, with the dwelling required to provide only basic protection and security in the night. Early windows, rather than being significant for daylight, were more important as a way to ensure that fresh air would reach the smoky interior caused by the central open fire. It was not until the 16th century that candle light and oil lamps became available, though an expensive source of interior lighting.

This is the vernacular architectural context of the great architectural masterpieces of past civilisations. Imagine the impact of coming from a modest Medieval dwelling, with a few shuttered unglazed openings and a smoky low ceiling interior, and entering the dramatically high, elegantly structured and beautifully light cathedral of Chartres. Such stark contrasts are difficult to imagine today.

1.2 Light and shade - revealing form

"Shadows have always been the brush work of the traditional architect."

Frank Lloyd Wright

In order to fully appreciate the design successes of ancient civilisations, it is necessary to understand the nature of the light in which their buildings are revealed. In southern climates dominated by clear skies and bright sun, the great architectural masterpieces exploited the intensity and directionality of light to great visual effect. It is particularly in the religious monuments of Ancient Egypt and Greece that we can witness the skill and subtlety with which intense light is controlled to inspire the

onlookers. These skills are based both on the wisdom of evolving vernacular explorations and on new knowledge of mathematics and solar geometry. As the quote from Frank Lloyd Wright suggests, the manipulation of light and shade in the clarity of the southern sky establishes the foundation of the rich and varied palette of daylighting design.

1.2.1 Egypt

The monumental stone temples set in the desert landscape manifest the deep understanding of the effects of sunlight and its diffusion from the vernacular tradition, but interpret this in innovative ways and on a grand scale. The role is now no longer simply one of protection and privacy, but of creating an atmosphere appropriate to the dedication of the gods or kings.

The pyramids were sheathed in white limestone and crowned by a pyramidion of gleaming precious metal that would catch the dawning sun before anything else. The obelisks too would be capped with gold to pick up the first and last light of the day. The monuments take advantage of the strong sunlight to reveal, through the contrast of light and shadow, three-dimensional form. Many of the structures of ancient Egypt are adorned with hieroglyphics – bas relief carving in stone – effectively revealed by the sharp directional light of this climate. It is not surprising that the ancient Egyptians above all worshipped the sun god Ra.

To illuminate the interior of the temples, light is introduced into the deep plans via clerestory elements consisting of narrow slots carved into stone slabs that act as giant stone *mashrabiya* (some are still intact). The intensity and location of light reinforced the axis of the main processional route. For example, in the Hypostyle Hall of the Great Temple of Ammon light is brought in through clerestory slotted slabs above the central axis supported on open lotus capitals (symbolic of the light) (Figure 1.10). In order to emphasise the expanse of the

Figure 1.10 Great Temple of Amun, Karnak, Egypt (1530–323 BC). Plan, section and detail of the hypostyle hall indicating how the intense sunlight is diffused and filtered into the hall via stone slits to provide a cool interior luminous environment. (Gregory Anderson)[17]

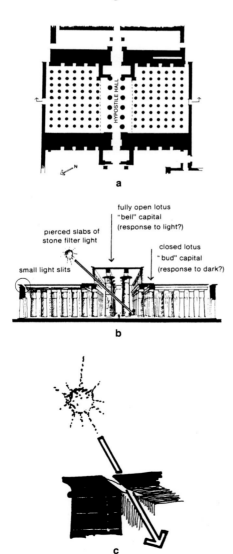

darker Hall in either direction, the lowered roof of which is supported by vast columns with closed lotus capitals (symbolic of the darkness), a relatively small amount of light is allowed to filter in through slots around the periphery of the enclosure. This light establishes a new artificial horizon of dimly lit stone, and can just be glimpsed through, and thus reveals, the forest of columns. The whole building form and sequence of spaces is manipulated to accentuate the processional movement from light to dark.

1.2.2 Greece

Further north, the Greek temples too demonstrate a clear understanding of solar geometry and lighting effects. The east orientation of the doorways enabled the low-angle morning sun – the first rejuvenating light of day – to penetrate deep into the interior and reveal the statues of deities within. Externally, the depth of the facade, with layers of closely spaced columns in front of solid stone walls, is revealed by the strong daylight (Figure 1.11). At a more detailed level, the sharply fluted channels on column shafts accentuate their curvature in direct sunlight through a changing rhythm of vertical lines of light and shade. The details of the triglyphs above set up a rapid rhythm synchronised with the broad beat of the colonnades. Entablatures and the bas relief carvings in the pediments, as with the hieroglyphics in Egypt, are revealed in the bright light. Furthermore, this light is dynamic, both in its intensity and in its directionality, imbuing the architecture and sculpture with life.

Daylight varies, being sometimes diffuse and soft, and at other times direct and sharp. The angle of incidence on the building facade changes constantly throughout the

Figure 1.11 The Parthenon, Athens. The sharp modelling of the fluted columns and the base relief of the frieze exploit the direct, oblique sunlight to animate the architecture. (John Donat, London)[18]

day and over the seasons. The whole effect, as seen from afar and close to, is of an architecture animated by the moving sun. Thus the ancient civilisations have demonstrated their knowledge of how the bright light of their climates can be manipulated to reveal architectural qualities beyond the purely formal or geometrical.

This heritage of light and shade, chiaroscuro, has been handed down through the ages and is embodied in the graphical technique of sciagraphy (the drawing of shadows). More recently, the use of a heliodon to project shadows of the sun's path on scale models is a technique that not only reveals the dynamic three-dimensional qualities of a design, but also allows the designer to predict sunlight penetration and patterns into the interior.

1.3 Structure and light - letting the light in

"Structure is the giver of light."

Louis Kahn[19]

Although the sunlight of Egypt and southern Greece has been expertly exploited to reveal the exterior form and surface modelling of ancient monuments, it is rarely allowed to enter the interior in any great quantity. This is not only to prevent excessive solar gains and thus potential overheating, but also to create a subdued dimly lit interior to invoke a mood of contemplative reverence. Furthermore, the structural limitations of the simple post and beam configuration made it impossible to attain significantly large openings.

If one considers more northerly latitudes, the winter days become shorter and colder, and more social activities take place indoors. In such a context, letting the daylight in becomes both important to reveal the interior and the activities as well as to allow any available sunlight in to warm the space.

1.3.1 Rome

The structural inventions of the Romans enabled larger openings and spaces to be created. The arch, vault and dome – together with the use of clear glazing – allowed greater column-free space to be flooded with light, in contrast to the narrow shafts of light in Greek or Egyptian post and beam architecture. The public Roman baths demonstrate most clearly the opening up of the architecture, with glazed arched windows facing the early afternoon sun for the hot baths, while cold baths were on the northerly side.

The use of large arched clerestory windows admitted daylight high up in the naves of the Basilicas, flooding the interior with daylight. The scale of such openings was not to reappear again until the Gothic period, when structural ingenuity enabled walls to virtually disappear from the interior into slender lines of structure.

In Roman society, much public activity took place in the large monumental buildings. This required effective daylighting strategies, through the inventive use of structure and glass as described above, to illuminate and warm the activities within.

Vitruvius was perhaps the first architectural author who studied daylight both qualitatively and quantitatively. He proposes the following explicit rule to assess whether an interior is well daylit:

"We must take care that all buildings are well lighted.... Hence we must apply the following test in this matter. On one side from which the light should be obtained let a line be stretched from the top of the wall that seems to obstruct the light to the point at which it ought to be introduced, and if a considerable space of open sky can be seen when one looks up above that line, there will be no obstruction to light in that situation."[20]

The angle of the view of the sky that Vitruvius is determining in this way is not dissimilar from the modern 'sky component' of daylight, although no clear criteria are set (e.g. what is "a considerable space of open sky"?). However, such functional considerations were not the only expression of daylight manipulation in Roman architecture.

The Pantheon in Rome (Figure 1.12) demonstrates a subtle and skilful use of light. A shaft of daylight enters high up in the spherically proportioned interior space through the oculus. As the sun moves, the sunpatch traces a path across the interior, producing strong shadows as well as scattering light diffusely into the vast interior to reveal all its full architectural plasticity.

However, it is not only the luminous quality of the interior that makes the Pantheon a particularly beautiful example of daylight design. It is the way that the approach and entry into the dome are modulated that sets the scene and leaves a deep impression, as Rasmussen describes:

"Coming to the Pantheon from the tangled network of streets outside, we experience it as the perfect expression

Figure 1.12 The Pantheon, Rome. Entering the vast open interior space from the colonnaded peristyle heightens the sensibilities, accentuated by bright sunlight from the oculus to draw the eye up to the heavens.

of peace and harmony. The ordinary scale of the houses just passed makes the peristyle, in comparison, seem overwhelmingly high with its gigantic columns disappearing into the twilight under the roof. As you enter the rotunda you are immediately aware of the mild light coming from a source high above you, three times as high as the ceiling in the peristyle. The dome does not seem to limit the space but rather to expand and raise it.... The circular opening at the summit of the dome forms the only connection with the outside world – not the noisy, casual world of the streets but with a still greater hemisphere, the celestial sky above."[21]

1.3.2 Gothic

The symbolism and imagery of light and dark were ideal vehicles for the expression of religious mysteries, and were used to inspire devotion. Structural ingenuity was exploited in Gothic architecture to provide a clear break with the past:

"The elimination of the massive wall structure and the frontality of Romanesque churches in favour of a lighter and more diaphanous structure with an emphasis on diagonal lines and views."[22]

The structural sophistication of Gothic architecture allowed the loads to be traced via ribs from the vaults and pointed arches laterally to point locations in the walls. Flying buttresses extending perpendicularly to the outside resist the outward forces of the vaults, whilst the vertical loads are transmitted down slender columns clustered in groups. The result is that the walls no longer have a load bearing role to play and become highly glazed with expanses of stained glass.

Abbot Suger (1081–1151) was patron of the work on the church of St-Denis near Paris, and it is in the ambulatory around the choir that the new Gothic architecture was embodied. The use of pointed arches and ribbed vaults on slender columns allowed space and light to flow freely, unconstrained by heavy supporting walls. Suger describes the result:

"...the whole church would shine with wonderful and uninterrupted light of most luminous windows, pervading the interior beauty."[23]

The association of God with light, and the use of stained glass to create a coloured and mysterious quality, are central themes in a Gothic architecture that creates a unique luminous experience.

The Gothic style culminates in the Perpendicular style, so called because of the rectilinear vertical stone panels, and the fan vault where the simple ribs have become an intricate filigree of decorative stone tracery over cone-like vaults. King's College Chapel is a magnificent example of the result (Figure 1.13). The weight of the fine stonework high up in the vaults is transferred out through the building's glassy walls, which appear too light and thin to support the load.

Figure 1.13 King's College Chapel, Cambridge. The refinement of Gothic structure using exterior buttresses allow the walls to be glazed to an unprecedented and awe-inspiring degree.
(A F Kersting, London)[24]

Enlightenment and reason

1.4.1 Defining light

By the time of the Renaissance two words were used to describe light: *Lux* and *Lumen*. Although the terms have quite distinct and clear definitions today, in the past they were defined more ambiguously as follows:

"Lux is the natural property of luminous bodies that imparts a motion similar to that of the body to which it belongs. This movement is its essence and does not depend on anything else intrinsic in the body. Lux was given its existence by the Creator at the act of creation of the world..."

"Lumen namely the illuminating light is the image of the light itself that is to say of lux, and its derivation is of a primary nature."[25]

In these terms Lux signifies the emotional response to light and cannot be measured. Lumen on the other hand refers to the light one can see and which thus can be measured, and is more tangible than Lux: Lumen and Lux could be construed to represent the rational and the emotional, or the scientific and the poetic aspects of light. It is evident that there was greater ambiguity between objectivity and subjectivity than today. This dualistic aspect of light has been a fundamental debate throughout the ages. Aristotle and Empedodes of Agrigentum in the 5th century BC considered light as the metaphysical link between the object and the soul, and that light could not be considered objectively. Conversely, Euclid in the 4th century BC established the geometry and physics of light, and thus the rational school of light, developed further by Ptolemy in the 1st century BC. The rationalist architects and builders based their approach on the empirical tradition, which, not until the late 19th century, developed into a mathematical approach.

During the Renaissance, light was no longer regarded as a symbol of the presence of God, as in Medieval times, but as an enhancement of the sense of life. Leonardo da Vinci, the scientist-artist, explored the nature of light, and commented:

"On studying the causes and motives of Nature, the observer is fascinated, above all, by light."[26]

The natural environment was, at this time, often used as the basis for linking and evoking emotional response to light. For example, the patron of the design of Pienza – the first ideal city of the Renaissance – Pope Pius II (Aeneas Silvius Piccolomini), had a love of nature, and describes the environment around Pienza in his early commentaries:

"Grassy fields spread beneath the chestnuts and are always shady except after the autumn frosts when leaves have fallen and the sun's rays penetrate the branches. Surely here if anywhere in the world, sweet shade and silvery springs and green grass and smiling meadows allure poets."[27]

In describing the design for the cathedral and palace at Pienza, Pius makes much reference to the luminous qualities. The cathedral's unusual orientation, entered from the north with the apse facing south, is described as follows:

"As you enter the middle door the entire church with its chapels and altars is visible and is remarkable for the clarity of light and the brilliance of the whole edifice."[28]

The palace is designed in accordance with rules regarding light and orientation set out by Alberti, who was a member of the Pope's staff. Alberti, the archetypal Renaissance man (scholar, author, mathematician and athlete), saw architecture as an intellectual discipline and a social art necessitating both artistic and mathematical skills.

The romantic and empirical descriptions of light in nature and architecture were followed, in the late Italian Renaissance, by Andrea Palladio's more mathematical rules to size openings. His texts expressed the need to strike a balance between wanting a "clarity of light" whilst not allowing in too much heat. Palladio proposes that:

"... the windows ought not to be wider than the fourth part of the breadth of the rooms, or narrower than the fifth, and to be made two squares and a sixth part of their breadth more in height."[24]

Despite such formulaic approaches in architectural treatises, the science of light was subordinate to the aesthetic effects of proportional systems. Even Palladio's villas, La

Rotunda and Villa Chiericato, do not conform to the rule and have more appropriate (lower) average daylight factors (3% to 4.5%) compared with the rule (7%). At this stage in architectural history the romantic and the pragmatic understandings of light in architecture seem to be approaching a degree of balance, with both having their impact on design.

The Renaissance revival of interest in visual harmony and proportion, inspired by classical architecture, results in a more subtle and clever manipulation of daylight, used to emphasise form and dramatise space. The modular and rhythmic arrangement of tall windows is typical and provide the appropriate 'universal light'.

1.5 Sculpting with light

1.5.1 Baroque

Baroque architecture – an architecture in which light is a central concern – is typified by a sculptural exuberance and dynamic spatial qualities, but has its basis in calmer classical and Renaissance traditions. The three-dimensional articulation of forms enables a more imaginative control of light introduced between the overlapping layers of the enclosure. Openings were no longer simple holes in simple walls, but perforated vaults behind which concealed light could reach other areas indirectly to illuminate the interior dramatically and mystically.

Unlike the Gothic cathedral, which can be seen as representing the light-filled heaven, the Baroque church represents the link between earth and heaven – between earthbound architecture and heavenly light. The use of mysterious light and indefinite spacial enclosure signifies transcendence.

Typically, the central space is illuminated by indirect light, and with the use of frescos, ornamental stucco and sculpture the boundaries of the space are indefinite. The illusions created in frescos and sculptures are either an extension of the real architecture and space (from a given point of view), or can be perceived as an opening cut into the vault with a view into the heavens (or a combination of both). These images and forms require a significant amount of light in order for their full impact to be visible. However, the source of light must be hidden from direct view so as to avoid glare – paintings will appear dark if there is a bright source within the visual field – and, particularly for sculptural decorations, light should fall on them obliquely to emphasise the three-dimensional forms with light and shade. As a way of creating the required

light conditions, external walls and the windows in them are hidden from view by surrounding the main space with an intervening light-filled mantle – a veil of divine light. In smaller buildings the relatively large windows are treated as ornament, with deep scalloped reveals, to minimise the attention that they draw to the walls.

The architect who had given the first expression of the Baroque style was Bernini, with the baldacchino at St Peter's, Rome (Figure 1.14). St Peter's expresses a 'carving out' of the interior space from the masonry enclosure. The

Figure 1.14 St Peter's, Rome (1607–c 1614). Parallel shafts of sunlight cut through the hazy incense-laden air to enter at the fulcrum of the building. (Sonia Halliday, Weston Turville / Jane Taylor)[30]

remaining thick walls and pochés between interior surface and external envelope allow daylight to be brought in secretly through numerous deeply recessed openings, which diffusely illuminate the interior.

Perhaps the pinnacle of Baroque design is the Trasparente (1721–32) in Toledo Cathedral by Narciso Tomé (Figure 1.15).

"The whole amazing sculptural group is dramatically lit by a golden light from a concealed source behind the visitor. When he turns to look back at this source he is confronted with a celestial vision contrived in the space created by the removal of one whole rib vault in the ambulatory. Above this Tomé constructed a large high dormer containing a window invisible from below. Carved angels frame the mouth of this bizarre opening..."[32]

Late Baroque, perhaps exemplified best by the Bavarian rococo church, extends the ideas of light and form to the interior reflectances and materiality of finishes used:

"At least as important is the way in which the white walls and pillars of the interior absorb this light, become immaterial and radiant. Light and matter fuse as stone and stucco are transformed into an ethereal substance... lets us forget the heaviness of the material with which it is built and helps to establish the sacred character of this architecture."[33]

Figure 1.15 The Trasparente, Toledo, Spain (1712–32). The voluptuous depth of Baroque architecture is revealed by concealed sources of daylight passing through the 'poches-vide' of the exuberant enclosure. (Clive Hicks, London)[31]

1.6 Painting with light

"This light describes the architectural envelope with objective precision, but at the same time gives a diaphanous immateriality to the wall surfaces, almost as if they were part of a watercolour and not an actual building."

Paolo Portoghesi[34]

1.6.1 Vermeer and the Dutch window

The Dutch domestic window since the Middle Ages is unique and particularly interesting. The tall windows in the gables of the narrow houses are largely the result of the physical conditions. The limited amount of land, and its cost due to land reclamation, resulted in the typical deep, tall and narrow gabled canal house. The openings are commonly divided into four frames, the lower two provided with external shutters and originally with no glass, and the upper two with fixed glazing (see Figure 1.16). This allowed shutters to be opened for ventilation and plenty of daylight, or closed with daylight entering high up and deeply into the plan. Later, both the lower and upper openings would be provided with shutters and made openable to increase the range of conditions that could be provided.

The beauty of 'simple' side lighting is most explicitly explored in the paintings of Vermeer (Figure 1.17). Many of the interiors he has painted exploit the strong three-dimensional qualities of sidelight to reveal faces, materials and textures. The paintings also show how the design of the opening can provide numerous configurations and daylight conditions in relation to the activity of the subject. In most cases the subject is concentrating on an activity, whether it is sewing, reading, playing the piano or weighing pearls. The sewer sits near the opening in a shaft of daylight, head bowed closely to the

Figure 1.16 The Dutch window: fixed panes above and shutters and inward-opening casement windows below.[35]

Figure 1.17 *The Music Lesson* (1662). Vermeer expresses the subtleties of meaning and control of daylight in a series of paintings that demonstrate the visual eloquence of the Dutch window.

detailed task to enable her to see accurately. The reader of the letter seems to have approached the window from the relative darkness of the room behind, and tilts the paper to the light. The pearl weigher has closed the lower shutters of the window to work in privacy, with daylight filtering through higher level glazing.

1.6.2 Architectural representations

"A superior manner of drawing is absolutely necessary, indeed, it is impossible not to admire the beauties, and almost magical effects in the architectural drawings of a Clerisseau, a Gandy or a Turner."[36]

In the above quote Sir John Soane recognises the importance of the ability to represent light in architecture. Soane is perhaps one of the great manipulators of daylight, as is evident from his designs, in particular his house – the Soane Museum (Figure 1.18). But it is perhaps most interesting to consider how Soane refined his daylighting skills through paintings. Soane consistently commissioned the artist J M Gandy to render his designs and represent their luminous environments. The relationship between the two men is well documented, but one can only speculate that Gandy's technical skills in representing the play of light and shadow have contributed to Soane's explorations and understanding of daylight in architecture.

Soane's long friendship and collaborations with J M W Turner are further indications of an exchange of understanding regarding their mutual interest in light.

At a practical level, the construction of the numerous rooflights in a variety of forms and configurations at the back of the Soane Museum evokes the impression of a scientific daylight laboratory. Even if this was not specifically the case, the resultant luminous conditions, represented in a tight sequence of spaces, offer fruitful insights into daylighting design. Soane's rational and artistic approach to design are discussed in his lectures, and summed up by him as follows:

Figure 1.18 The Breakfast Room (1812), Sir John Soane's Museum, London. Soane's 'poetry of architecture' is encapsulated in this small room through the careful placing of light slots (making the dome appear to float), central lantern light (focusing the space) and numerous circular convex mirrors (giving a sense of sparkle and brightness). (Courtesy of the Trustees of Sir John Soane's Museum).

"Civil Architecture ... is both essential and ornamental, it is partly an Art and partly a Science ... we soon perceive that its leading principles are founded on the immediate laws of Nature and Truth ... we shall then do justice to the wonderful powers of Architecture and place it in the first rank of refined Art and exalted Science."[37]

1.7 Technology comes of age

1.7.1 Light structures

The Industrial Revolution brought with it the development of new materials and technologies that were to reduce greatly the traditional constraints on opening size. The early railway sheds of the 1830s and 1840s revealed the potential of iron structures to liberate the space of cumbersome loadbearing walls. In an even more striking way than the Gothic cathedrals, space and light could flow freely. The increased sizes of clearglass panels reduced the need for close framing and increased transparency. With iron and glass railway sheds also came the more refined greenhouses and arcades (Figures 1.19 and 1.20). An extension and elaboration of this tradition is Joseph Paxton's astonishing structure, the Crystal Palace, to house the Great Exhibition of 1851.

The transparency that iron and glass offered was appropriate for buildings exhibiting growing plants or pro-ducts. Daylight is maximised to encourage photosynthesis in tropical flora, and for the display of goods, whilst being sheltered from the vagaries of the north European climate. The improved microclimate – protection from rain, wind and the cold – is conducive to shopping, whilst the thin iron and glass roof structure limits the reduction of daylight.

1.7.2 Labrouste

An early application of iron in a monumental public building is at Labrouste's libraries in Paris (Bibliothèque Ste-Geneviève, 1843–50, and Bibliothèque Nationale, 1859–68). In both, the slender iron columns and arches – representing industrialised society in a manner both rational and poetic – are revealed by daylight. At Ste-Geneviève side light enters through tall windows deeply into the reading space, whereas at the Bibliothèque Nationale (Figure 1.21) daylight enters through oculi in the domed ceilings and through high, north-facing windows. The rooflights provide an even level of daylight across the plan, but more importantly they light the structure that cuts through the space as slender slivers of light supporting an almost tent-like ceiling.

Figure 1.19 Iron and glass technologies were exploited to create buildings such as the Kew Palm House, London, for cultivating plants.

Figure 1.20 Commercial gallerias such as St Hubert, Brussels, also exploited contemporary structural and glazing technologies.

Figure 1.21 Main reading room, Bibliothèque National, Paris (1859–68), by Labrouste. The graceful iron structure is revealed in daylight by the circular rooflights in the domes, contrasting with the heavy masonry of the enclosing walls. (James Austin, Cambridge)[38]

In the US, the structural possibilities of iron, together with the arrival of the passenger elevator, gave birth to the skyscraper in Chicago following the devastating fire in 1871 and the need to rebuild quickly. Facades were fixed to the skeletal frame and no longer had to perform a structural function. This allowed a freedom of facade composition and window design.

1.7.3   Art Nouveau

"For Art Nouveau, light is an indispensable nourishment and a congenial instrument of wonder".[39] The form of Art Nouveau buildings was influenced by the emphasis on the use of 'new' materials such as iron and glass of the preceding period. The use of iron made it possible to achieve a lighter structure, and when used to express the flow of forces, gained an organic quality. Interior artificial light fittings took on the same motifs and became integrated into the themes. An example is Horta's Waucquez

Figure 1.22 Art Nouveau designers revelled in the design of new artificial lighting fixtures as an integral element of the space (Waucquez Department Store, now a museum of comic art, Brussels).

Department Store in Brussels (Figure 1.22). Here light floods down the central lightwell, giving the visitor a luminous focus. This luminosity is increased and spread deeper into the plan by translucent floors immediately around the stairs, which help to soften the contrast. This luminous core to Horta's buildings, excavated within the deep narrow sites of urban Brussels, overturns the traditional relationships. The centre is no longer dark and calm, but bright and dynamic with colours, celebrating the interior life of the buildings.

1.8 Man versus nature

1.8.1 Artificial lighting technology

Daylight was the predominant source of light until the introduction of gas lighting in the early 1800s and electric lighting by 1900. The earliest form of artificial light after the open fire was the candle and oil lamps of the 16th century. Even by the 18th century the cheap tallow candles were crude and irritable to the eyes, whilst the cleaner beeswax candles were reserved for the wealthy. It was in 1783 that Argand invented an oil lamp with a wick and glass funnel to improve control and intensity, even though fumes and heat were still generated, requiring good ventilation.

Gas lighting offered superior brightness, and was initially used for street lighting in London (1806), Paris (1819) and Berlin (1826). Mains gas lighting was available in most urban homes by the 1840s, although the smell of fumes demanded better ventilation. Up to this date the architecture was still designed to be substantially daylit. The cost and associated environmental problems and risks did not make gaslight a viable alternative.

The arrival of electric lighting in 1878 promised a bright and clean light source. By 1900 electricity supply was an accepted fact of urban life. This brought with it not only electric light, but also sewing machines (1889), fans (1891), vacuum cleaners (1901), washing machines (1909), etc.

The 19th century art of lighting began to give way to the 20th century science of lighting. In 1904 the Commission International d'Eclairage (International Commission for Lighting) was founded and began to propose standards and develop the science of photometry. It also spawned the lighting engineer.

1.8.2 Architectural repercussions

"The functionalist statute ... developed and exalted the theme of transparency and light, but tended to strip it of its dialectic nature and of any possible metaphysical connotations. The rational and Cartesian light of the functionalists is not born from darkness, and has no need for its opposite in order to exist and assert itself..."

Paolo Portoghesi[40]

The combination of technologies such as the invention of air conditioning by Willis Carrier in 1928, and the fluorescent tube by GEC in 1938 (reducing heat and glare), had a major influence on architectural possibilities. Building design could now be totally independent of the exterior climate – artificially lit, heated and cooled. The use of steel and glass meant there were also no constraints on building form or facade opening design. The immediate result was the construction of high-rise, deep-plan, tower blocks still prevalent in our cities today.

Walter Gropius with the curtain wall glazing at the Fagus Factory (1911–12) (Figure 1.23) and Mies van der Rohe with his proposal for a glass skyscraper in 1920 (Figure 1.24) pre-empted architectural possibilities that later technological developments of environmental control would make feasible. The fully glazed transparent aesthetic has had a powerful impact on modern architecture. However, the advantages of high daylight levels around the perimeter of the building were offset by increased glare, excessive solar gains and heat losses, and a lack of privacy – "leaving the inmate defenceless in a glass cage" (F L Wright). Air conditioning could begin to address the thermal problems, but at a great cost in terms of energy consumption. Privacy and glare could be improved with blinds, to the detriment

Figure 1.23 Fagus Factory, Alfeld-an-der-Leine, The Netherlands (1911–12), by Gropius. The curtain walling and glazed corner staircases of this building, representing transparency and movement, were to influence the international Modern Style.

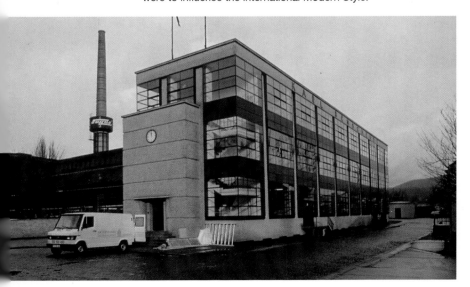

of the architectural intention of transparency. The darker, deep-plan interiors could be lit with fluorescent lighting, although again requiring more energy use.

1.8.3 The post-fluorescent era

The lack of the need to feel concerned about daylight and sunlight for several generations of architects, by reliance on technology and cheap fuel to resolve lighting (and thermal) issues, has meant that much of the art of daylight design has been lost.

In the meantime, scientific progress has allowed us to describe, measure and predict light with great certainty. Being able to determine the illuminance on a given plane tells us little about the emotional content of a luminous environment. An interior that achieves the required lux levels on the working plane will not necessarily be a beautiful space. However, applying scientific analysis to buildings that we appreciate, know or can experience, will give us the experience with which to begin to understand what makes a space that we find restful, stimulating, inviting, gloomy, bright, etc.

Figure 1.24 Design for a skyscraper (1920–1), by Mies van der Rohe. An ominous preview of modern, not-so-transparent, highly glazed office buildings, with no regard for the subtle manipulations of daylight or control of sunlight.[41]

1.9 Modern masters of light

"It is perhaps just because of this long fasting, of this abstinence from the search for not absolutely transparent light effects, that one of the most obvious symptoms of architecture after the Second World War ... was precisely a return to the shadow-light dialectic and to the quest for lighting effects capable of restoring a lost dimension to architecture."

Paolo Portoghesi[42]

1.9.1 Asplund

A number of recent architects have built on the traditions of vernacular and high architecture. A response to climatic and cultural context, with a real and explicit awareness of the power and meaning of natural light, has made them stand out as vernacular or romantic modernists. Gunnar Asplund (1885–1940) in particular had flirtations with the International Modern style but made highly original contributions to it. Asplund's "romantic primitivism" as expressed in the Woodland Chapel (1915) – a combination of a temple and a woodland cabin – is characteristic of the search for modern simplicity rooted in a classic tradition. His major works are the Stockholm City Library (1920) and the Law Courts Annex in Göteborg (1937) (Figure 1.25).

The Law Courts building is particularly expertly lit through daylight entering and the atrium / Great Hall,

Figure 1.25 Goteborg Town Hall, Sweden, by Asplund. The combination of sidelight and toplight in this atrium space creates a sense of dynamic that is reinforced by the location of the vertical circulation route between the two sources of daylight.

both through a south-facing glazed facade looking into the courtyard, and a south-facing rooflight. The combination of sidelight and toplight creates a dynamic quality, changing the lighting mood as you pass from form-giving lateral light to a more 'centred' nature of vertical light in the depth of the Great Hall. The two high-level walkways also have strongly differentiated light intensities, the interior one being significantly lighter than the other. The shape of the rooflight ensures that even low-angle winter sun can be reflected down off its soffit into the hall. This set of manipulations results in a formally simple architecture being animated and modulated by daylight.

1.9.2 Aalto

The Finnish architect Alvar Aalto (1898–1976) used daylighting as a primary design consideration for most of his buildings. Recurring features are the lightscoops and conical rooflights found for example in the Library in Rovaniemi (1965–68) and the Museum at Aalberg (1969–73). The conical rooflights provide focal points and illuminate the horizontal planes (such as reception desks or circulation areas, reinforcing the sense of movement). Sunlight cannot enter the conical rooflights because their depth is sufficient to intercept and diffuse the low sun angles of Finland. The scoops tend to reflect more light vertically onto the walls of exhibition or book stack areas. All these elements have white finishes to ensure that light is reflected down, and they are designed to intercept unwanted low-angle sunlight whilst being open to the predominantly overcast skies of Finland.

1.9.3 Le Corbusier

Le Corbusier completed two religious buildings in France that express a keen engagement with the manipulation of daylight in favour of concerns for the rational principles of the modern movement. In particular the pilgrimage chapel at Ronchamp (1950–55) (Figure 1.26) uses voluptuous flowing concrete forms to scoop light to focal points such as the altars, or introduce daylight through an array of deeply recessed openings in the key wall.

At the Dominican Convent of La Tourette, Le Corbusier's austere concrete volumes are pierced by *canons de lumière*, trapezoidal rooflights and an array of light slots. The three *canons de lumière* face in three different directions and are painted three different colours on the interior conical surfaces. The colours 'paint' the three pools of light, defining the private prayer areas in the crypt. The seven trapezoidal rooflights face south and

bring columns of light into the sacristy. The light slots cut into the walls of the box of the chapel have reveals painted red, white, yellow or green to bring colour into the otherwise grey interior. The light slots are used to 'draw' key geometrical lines of light to define the space, often located near perpendicular walls to reveal the texture of the surface and the directional qualities of the light. The contrast between the simple and rough concrete containers of space and the flowing coloured richness of the light emphasises the difference between earth-bound mortality and heavenly eternity.

Figure 1.26 Light filtering through deep reveals and coloured glass in Le Corbusier's light wall at Notre-Dame-du-Haut, Ronchamps, France (1950–5).

1.10 Summary

Through history the role of daylight has been one to mediate between the romantic and prosaic. It has often been considered either solely in terms of aesthetic purposes or to meet functional requirements. In fact, light always provides both. In a church the light is subdued and inspirational, but sufficient to read by, once the eyes have adapted. Conversely, the need for bright conditions in a kitchen is integral to the character such a space tends to have. Daylight is to be used both as an aesthetic medium and a pragmatic technique to reduce our dependence on electric light. The integration and complementarity of daylight quality and quantity gives the potential to achieve sustainable and beautiful architecture.

References

1. Le Corbusier, *Towards a New Architecture* (Oxford: Butterworth Architecture, 1989), p. 29 (translated by F Etchells; first published as *Vers une Architecture* by Editions Crés, 1923).

2. R Buckminster Fuller, *Critical Path* (New York: St Martins Press, 1981), p. 62.

3. H Fathy, *Natural Energy and Vernacular Architecture* (Chicago/London: University of Chicago Press, 1986), p. xx (Preface).

4. D Watkin, *A History of Western Architecture* (London: Barrie & Jenkins, 1986), p. 139.

5. From H A Bernatzik, *Südsee: Ein Reisebuch*, in B Rudofsky, *Architecture Without Architects: An Introduction to Non-Pedigreed Architecture* (New York: Museum of Modern Art, 1964), picture no. 115.

6. *Ibid*, picture no. 110.

7. Fathy, *Natural Energy and Vernacular Architecture*, p. 99.

8. *Ibid*, p. 101.

9. *Ibid*, p. 98.

10. *Ibid*, p. 143.

11. K Butti and J Perlin, *A Golden Thread: Five Hundred Years of Solar Architecture and Technology* (London/Boston: Marion Boyers, 1980), pp. 3–4.

12. *Ibid*, p. 5.

13. K Butti and J Perlin, *A Golden Thread*, p. 5.

14. M H Morgan (translator), *Vitruvius: Ten Books on Architecture* (New York: Dover Publications, New York, 1960), p. 170.

15. *Ibid*, p. 181.

16. Butti and Perlin, *A Golden Thread*, p. 19.

17. F Moore, *Concepts and Practice of Architectural Daylighting* (New York: Van Nostrand Reinhold, 1991), p. 4.

18. D Watkin, *A History of Western Architecture*, (London: Barrie & Jenkins, 1986) pp. 23.

19. L Kahn, "The room, the street and human agreement", *AIA Journal*, vol. 56, no. 3, September 1971, pp. 33–34.

21. S E Rasmussen, *Experiencing Architecture* (Cambridge, MA: MIT Press, 1964), pp. 192–193.

22. D Watkin, *A History of Western Architecture*, p. 126.

23. *Ibid*, p. 127.

24. *Ibid*, p. 155.

25. V Ronchi, *The Nature of Light* (London: Heinemann, 1970), p. 67.

26. I A Richter, *The Literary Works of Leonardo da Vinci* (Oxford: Oxford University Press, 1939), vol. 1, pp. 12–14.

27. Aeneas Silvius Piccolomini, *The Commentaries, Secret Memoirs of a Renaissance Pope* (translated F A Cragg) (London: George Allen & Unwin, 1960), Book IX, p. 270.

28. *Ibid*, p. 279.

29. Andrea Palladio, *Four Books on Architecture* (translated Ware), 1738 (London: Dover Publications, 1965), p. 30.

30. D Watkin, *A History of Western Architecture*, p. 213.

31. *Ibid*, p. 304.

32. *Ibid*, p. 307.

33. K Harries, *The Bavarian Rococo Church: Between Faith and Aestheticism* (New Haven/London: Yale University Press, 1983), p. 73.

34. P Portoghesi, "Light and modern architecture" in Y Futagawa (ed), *Light & Space* 1 (Tokyo: ADA, Edita Tokyo Co. Ltd, 1994), p. 6.

35. Rasmussen, *Experiencing Architecture*, p. 201.

36. J Soane, *Lectures on Architecture* (ed A T Bolton) (London: Sir John Soane Museum Publications, 1929), Lecture V, p. 88.

37. *Ibid*, p. 17.

38. D Watkin, *A History of Western Architecture* (London: Barrie & Jenkins, 1986), p. 385.

39. Portoghesi, "Light and modern architecture", pp. 7–8.

40. *Ibid*, p. 9.

41. D Watkin, *A History of Western Architecture*, p. 522.

42. Portoghesi, "Light and modern architecture", p. 13.

Part 2
Daylighting design

2 Climate and context

The shading and balconies in this street in Valletta, Malta, are a rich response to the Mediterranean climate

Climate and context

We have seen in Chapter 1 how traditional responses to climate have recognised the different availability of daylight – the small shuttered windows of southern Europe, defensive against the strong sunlight, and the larger windows found in central and northern Europe, in spite of the colder weather. Furthermore, in many countries we find evidence of legislation and planning codes, which protect the site against the obstruction of daylight by other buildings.

It is, then, the availability of daylight at the site that constitutes our starting point. Most designers will already have an intuitive understanding of the local daylight conditions, gained from personal experience both as a professional and also from living in the same environment. It may not always be necessary, therefore, to undertake a scientific study for this part of the site analysis. But for the sake of offering a comprehensive description, this handbook covers this part of the design process also, although in a rather summary manner.

The objective of this chapter is to help designers to understand the nature of daylight as a resource and to establish a basis for their daylighting design. By comparing the actual daylight resource available with the requirements set by the building programme, one can establish daylighting targets for the building project in question, and also evaluate different strategies for satisfying these targets. The available daylight resource is primarily a function of the overall daylight climate, which to some degree is modified by local conditions.

The daylight requirements in the building are related to the illumination levels recommended for the different visual tasks, according to the building's function (2.2) and the fraction of the occupied time that these levels can be achieved by daylight. In some cases, building codes and other regulations will in addition set rigid specifications for daylight in rooms for longer occupancy periods. For example, a reasonable target for a school classroom might be that an illuminance level of 200 lux is exceeded for 80% of the school year during the occupied day.

2.1 The sun and sky

The available daylight that can replace artificial lighting is both direct sunlight and diffuse light from the sky. The regional or overall daylight climate at a location is determined primarily by latitude and cloudiness. Atmospheric conditions other than clouds, such as gaseous and particulate pollution, and reflected daylight from the ground, will also have some impact on the daylight climate at a more local scale.

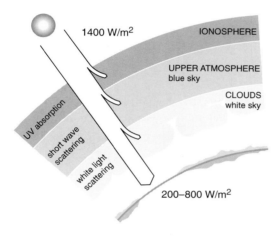

Figure 2.1 The scattering of sunlight by the atmosphere.

Sunlight reaching the thin layer of atmosphere is scattered – that is to say, it is redirected by the molecules that constitute air, as well as by water and dust particles suspended in the air (Figure 2.1). The former leads to the blue sky colour, since the blue end of the spectrum (shorter wavelength) is scattered more than the red end. The larger water and dust particles (and, in urban areas, pollution) scatter all wavelengths and give rise to the white overcast sky. Light is also reflected from clouds, and the size of water particle may affect the degree of reflection, hence giving rise to grey clouds.

The geometry of the sun path relative to the earth is known precisely, and this enables us to predict the solar altitude (vertical angle from the horizon) and the azimuth (direction relative to north or south). The latitude, longitude, date and time of day are the data needed to calculate the exact position of the sun in the sky. For convenience the geometry is often represented on a sunpath diagram, of which two types are available: the cylindrical projection and the stereographic projection.

The cylindrical projection (Figure 2.2) represents the azimuth, or direction of the sun from the point of interest, on a horizontal axis, and the vertical altitude of the sun on a vertical axis. The line at the top of the graph represents an altitude of 90°, i.e. directly overhead, and the line at the bottom represents the horizon. The resulting view can be likened to a panoramic photo.

The stereographic projection (Figure 2.3) represents the whole hemispherical sky dome as a circular disk with its

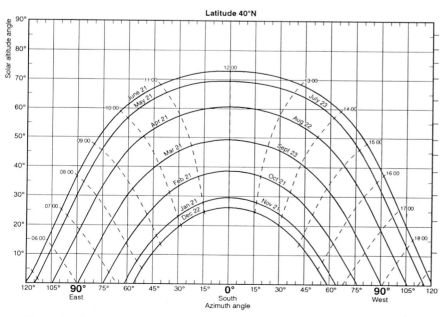

Figure 2.2 Cylindrical projection of the sun's path across the sky for 40° latitude.[1]

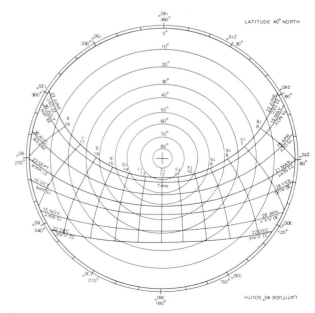

Figure 2.3 Stereographic projection of the sun's path across the sky for 40° latitude. (BRE).

Figure 2.4 Papageorgiou Hospital, Thessaloniki, Greece. Patients' view through a hospital window showing low winter sunpath and shading function of overhangs for summer sunpath. Sunlight reflected from the ground and other parts of the building makes a useful contribution to the daylighting of the room.

centre corresponding to the zenith, i.e. vertically overhead, and its circumference representing the horizon. The resulting view can be likened to a 180° fish-eye photograph taken by a photographer lying on the ground looking upwards.

For a given latitude, the path the sun takes across the sky can be represented as a series of curved lines for different times of the year. Lines at right angles to these curves are the hour lines. Note that the hour referred to is solar time, not civil time or local time. Solar time, for a given location, is defined so that 12 noon occurs precisely when the sun is at its highest elevation. Usually, solar time and local time do not differ by more than about one hour, although summer 'daylight saving' shifts the local time a further hour.

Not only can the charts be used to give the sun's azimuth and altitude at any time of day and year, but terrain, neighbouring buildings or vegetation can be drawn on the chart to indicate when the sun will be obstructed.

2.1.1 Sunny and cloudy skies

The relative position of the sun is of vital importance for the prediction of daylight and passive solar heat gain through apertures. The same information is also needed to design solar shading devices (7.2). However, it's important to stress that for predominantly cloudy climates, as found in central and northern Europe, daylighting design is based upon cloudy overcast sky conditions – i.e. the diffuse sky. A patch of sunlight is between 5 and 10 times brighter than the same surface illuminated from the diffuse sky, and special elements

have to be used to re-distribute and use this light (9.1). Where sunny skies are relatively uncommon, this provision is not justified.

Only in predominantly sunny climates, as in southern Europe, does the direct sunlight make a significant contribution to indoor illumination by daylight. In these cases, however, it is still usual to prevent the ingress of direct sunlight (Figure 2.4) and rely upon reflecting surfaces to redistribute the luminous energy. In the simplest form, this surface will be the ground outside the building. This issue is discussed in section 7.2 on shading design, and more sophisticated approaches are discussed in Chapter 9 on advanced daylighting of systems.

For cloudy climates, the solar geometry is mainly important in the avoidance of glare and the quantification of solar gains.

As a general rule, the daylight levels, from both sunlight and diffuse skylight, increase with solar altitude (Figure 2.5). This can be readily explained by considering the

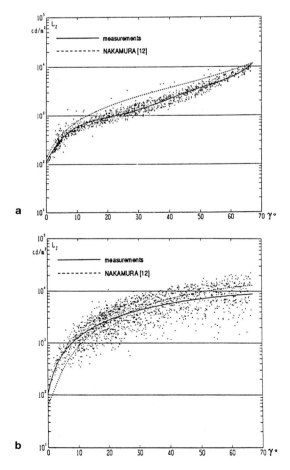

Figure 2.5 Sky zenith luminance as a function of the sun's altitude for (a) the diffuse sky, and (b) the clear sky.[2]

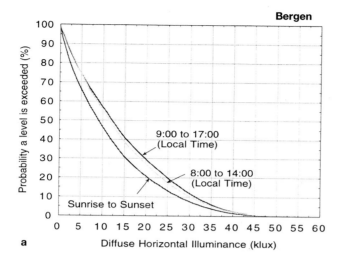

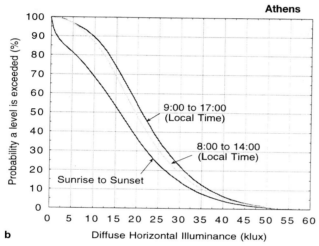

Figure 2.6 Daylight hours over the year for (a) a high latitude (Bergen, Norway) and (b) a low latitude (Athens, Greece) location. These curves are described in more detail in 12.2.[3]

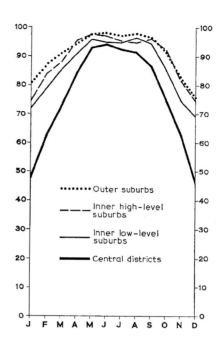

Figure 2.7 Monthly bright sunshine for inner London as a percentage of that outside London.[4]

distance that the sunbeams have to travel through the atmosphere. A low latitude location will therefore experience more daylight than at high latitudes, both in terms of momentary illumination levels, because the sun is higher in the sky, and in terms of useful daylight hours during a standard daily working period. At high latitudes, the short winter days are compensated by longer summer days. Actually, the annual sum of daylight hours is the same all over the globe, but with a standard working day, typically from 0900 to 1800, one cannot realise any savings from midnight sun (Figure 2.6)!

Daylight levels are to some degree modified by the local atmospheric conditions. A rural site will have cleaner air than an urban site (Figure 2.7). Pollution and humidity in the air will tend to diffuse the direct beam sunlight somewhat, which will raise the skylight illumination levels at the expense of the beam sunlight intensity.

The regional meteorological conditions at the location will determine the cloudiness and hence the statistical distribution of clear, overcast and intermediate days (or hours). Cloudiness is often indicated as an index between 0 and 1 (or 0 and 8, or 0 and 10), representing the fraction of the sky that is covered by cloud. Another indication of cloudiness is sunshine probability, the statistical chance for sunlight as a percentage of the theoretical maximum, the latter being given by the number of hours that the sun is above the unobstructed horizon. This is often referred to simply as 'sunshine hours' and is often available from the same meteorological sources as temperature.

As a general rule, coastal regions have much cloudier skies than inland regions, except those at high latitude. High latitude locations also tend to have more clouds and hence less relative sunshine hours (Figure 2.8).

The sky luminance (brightness) distribution for a fully overcast sky is quite different from that for a clear sky. An overcast sky is brightest in the zenith area, i.e. immediately overhead, and darkest along the horizon. In a location with a predominantly cloudy climate, as in northern Europe and many coastal areas elsewhere, toplighting in the form of roof apertures will thus give much more daylight than ordinary windows in walls. This effect is further emphasised by the fact that, in most cases, daylight is required for horizontal task surfaces. In fact, an

Figure 2.8 Relative sunshine duration February to April.[5]

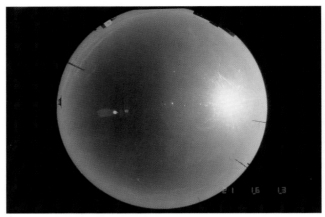

a

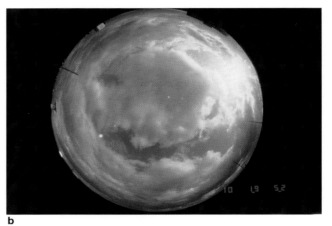

b

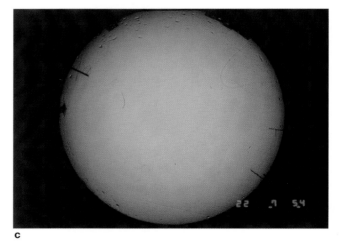

c

Figure 2.9 Skyvault photos illustrating the brightness distribution for (a) clear sky, (b) partly overcast sky, and (c) fully overcast sky.

area of roof aperture will typically yield 2 to 3 times more daylight on a horizontal surface than the same area of side window under overcast conditions (4.3).

During clear sky conditions the sky is brightest in a zone around the sun, but also quite bright along the horizon. The darkest part of the sky is opposite the sun in the same vertical plane, in a direction 90° away from the sun's position. The luminance distribution of a partly cloudy sky can be anything in between these two extreme situations (Figure 2.9). A more detailed mathematical treatment of sky types is given in reference 6.

The optimal placement and sizing of daylight apertures is thus sensitive to the type of climate at the location, and for the designer it is very important to realise what type of sky is predominant for the site in question (Figure 2.10).

Vernacular architecture will often demonstrate a very sensible adaptation to local climatic conditions (Figure 2.11). Traditional choice of window designs may offer a useful indication of the local daylight climate, which can be exploited in new buildings.

Long-term measurements of daylight or cloudiness can be used to produce statistical daylight availability data, which combine the effect of latitude and cloudiness, as already shown in Figure 2.6. This shows the fraction of the day between sunrise and sunset for which the illuminance from the diffuse sky exceeds the values

shown on the x axis. This type of data is necessary for the evaluation of the energy saving potential from daylight in buildings, as described in 8.2 and 12.2. More comprehensive data, for specific sites, are given in 12.2.

There are some interesting problems in using these data. For more southern climates, the major difficulty in

a

b

Figure 2.10 Examples of adaptation to local daylight conditions in (a) a sunny climate (architect's office, Athens, Greece), and (b) in a cloudy climate (law courts, Goteborg, Sweden).

establishing daylighting targets is the role direct sunlight is intended to play. Sunlight can either be admitted directly into buildings or, more commonly, exploited by letting reflected sunlight from exterior surfaces enter the building (4.3). The use of these diffuse curves will tend to underestimate the potential for daylight.

Precise sky luminance measurement has shown, too, that even the diffuse sky is not symmetric about a vertical

axis; a south-facing window collects significantly more useful light than a north-facing window. This can be accounted for with reasonable accuracy by the orientation factors published in the *European Daylighting Atlas*,[7] an example of which is also shown in Table 2.1 and described in more detail in 12.2.

The short daylight period and the rather low illumination levels at high latitudes will at some times of the year be partially compensated by reflections from snow on the ground. A simplified explanation is that daylight is reflected back onto the clouds and thus gives more daylight available indoors, because the sky will appear brighter. The snow-covered ground is of course also

Figure 2.11 Vernacular architecture illustrates response to climate in window design in (a) small defensive shuttered windows in Malta, and (b) large areas of glazing in Flemish architecture in Belgium.

a

b

Table 2.1 Orientation factors for Bergen, Norway. These factors account for the increased availability of daylight from the southern sky during typical occupancy periods.[8]

	North	East	South	West
Day	1.14	1.39	1.45	1.25
9–17	1.04	1.27	1.60	1.15
8–14	1.05	1.50	1.61	1.05

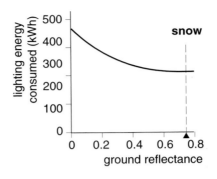

Figure 2.13 LT (8.2) generated curve of lighting energy as a function of ground reflectance. Reflectance of snow indicated.

much brighter than grass or nearby buildings, and reflected light from the ground is 'seen' by the ceilings in rooms adjacent to snowfields (Figures 2.12 and 2.13).

2.1.2 The daylight microclimate

Local conditions at the building site may modify the overall daylight climate. Some sites are more exposed to pollution (e.g. from nearby industry), while others are more prone to fog and haze due to local variations in air temperature, wind, and humidity. But the most important impact on daylight microclimate is due to obstructions, which shade the site from part of the sky and the direct sun at certain periods of day and year.

If these obstructions are quite far away (Figure 2.14) – i.e. mountains or other terrain formations – the shading will be apparent over the whole site, and the impact on the daylight conditions can be analysed for one representative point on the site only. If the obstructions are

close to the site – other buildings, vegetation, etc – overshading may affect only parts of the site, and solar conditions have to be studied in more detail.

The impact of obstructions on the sunlight and daylight conditions at a site is discussed in more detail in Chapter 3 and can be analysed with a range of different methods, which are described in Part 3.

It is important to realise that in principle one can analyse only one parameter at a time: either the variations of

a

b

Figure 2.14 (a) Local obstructions at the Queen's Building, Leicester, UK and (b) surrounding mountains at the Collège La Vanoise, Modane, France.

Figure 2.12 View from interior to snow-covered surfaces, which greatly increase reflected components.

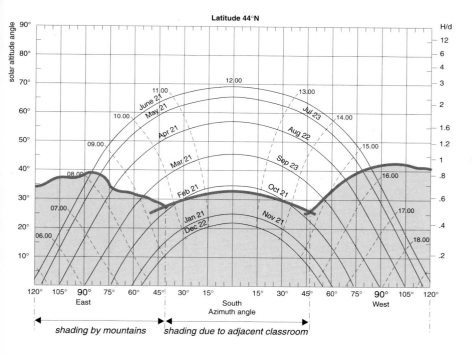

Figure 2.15 Sky view from the Collège La Vanoise (for specific reference point) shown on cylindrical sunpath.

solar availability over the year at one specific point, or the variations across a larger area (site) at one specific date and time. For the first type of analysis, a sunpath diagram for the point in question can be used to show for what periods of the year and the day the sun will be obstructed (Figure 2.15). The geometry of the obstructions can be recorded with surveying techniques, fisheye photography or analysis of maps.

For the second type of analysis, model studies with an artificial sun are the most powerful tool. An alternative is to use a solar shading calculation procedure in a CAD program. These methods are discussed in more detail in Chapter 11.

Another important point is that obstructions not only shade direct sun, they will typically also obstruct parts of the sky vault and thus reduce the diffuse skylight illumination levels as well. In standard daylight analysis it is customary to assume that obstructions have a brightness

(luminance) that is only 1/10 of the brightness of that part of the sky which is obstructed. The daylight contribution from the shaded part of the sky vault is thus reduced by 90%.

Obstructing the diffuse sky will reduce the availability of daylight within the building and thus require longer periods of artificial illumination, increasing energy consumption. This is shown in Figure 2.16, derived from the LT method (8.2.7).

However, adjacent buildings may make positive contributions as well. Consider a north-facing window, looking out onto a light coloured south-facing wall that is being strongly illuminated by the sun. This surface may be brighter than the unobstructed blue sky. Unfortunately these situations are highly specific to the time of day and particular geometry of the buildings, and so no generalised way exists to describe this luminous microclimate. However, modelling techniques (4.4) can be used to evaluate particular cases.

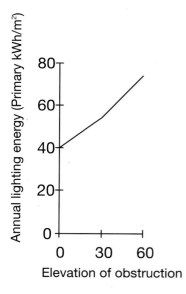

Figure 2.16 Lighting energy required for typical office space in the UK as a function of elevation of obstruction from centre of the window. Calculated by the LT method (8.2).

2.2 Daylight strategies and building function

Daylight can replace artificial illumination, and thus save electricity, and this is one of the main concerns of this handbook. But daylighting is much more than mere illumination; the changing nature of the light, the spectral composition and the visual communication associated with windows are in many respects equally important. When setting our benchmarks for energy savings through daylighting in a building project, it must not be forgotten that the amenity value of daylight also plays an important role. Many countries have introduced building codes or other forms of regulation that are intended to ensure that buildings will provide access to daylight and views out for the occupants. One will also find cultural differences in people's attitude towards daylight and the importance of windows, very often related to climatic characteristics in different regions.

The daylight available at the site – relating to climate and site properties – and the functions that require illumination in the building provide the starting point for analysing where daylighting is a viable option, and to what degree one can expect to replace artificial lighting with daylight. The first question to be raised is how good is the matching of the best daylight hours to the period when illumination is required? Daylighting is obviously not a very successful strategy for a night club! Fortunately, a substantial proportion of buildings are primarily used during daylight hours, particularly in relation to work, education and culture, including such buildings as offices (Figure 2.17), schools (Figure 2.18), shops, factories and exhibition halls.

Dwellings and other buildings for night occupancy, such as hotels and hospitals, constitute a special group in this respect. In some cases these buildings are not primarily occupied during the daylit period (hotels for example), so that daylighting is of lesser value. In other cases, such as hospital wards, the rooms are also used during the day, thus lending themselves well to electricity-saving measures through daylighting, over and above the vital role played by daylighting in the well-being of the patients.

Certain buildings have very specific requirements for illumination that may tempt designers to use artificial light rather than attempt to control the widely varying daylight source. A good example of this is the museum or art gallery. Although the special qualities of natural light are often recognised, the damaging effects of ultraviolet and the problems of overheating are often sufficient to dissuade the client from considering daylight. Certainly the technical solutions necessary to balance the visual needs of visitors with the requirements for conserving the exhibits are sophisticated, tending to demand state-of-the-art analysis and technology (Figure 2.19). But a growing trend is to recognise that the need for closely controlled environments has been overstated, and new approaches to light exposure management mean that daylighting is likely to become more acceptable (Figure 2.20).

Figure 2.17 The Domino Haus office building in Reutlingen, Germany. The use of daylight in the workplace makes a valuable contribution to energy conservation and to the well-being of occupants.

Figure 2.18 Collège de la Terre Sainte, Coppet, Switzerland. Designing for daylight in Swiss schools has a long history, reflecting concern for health and good visual conditions.

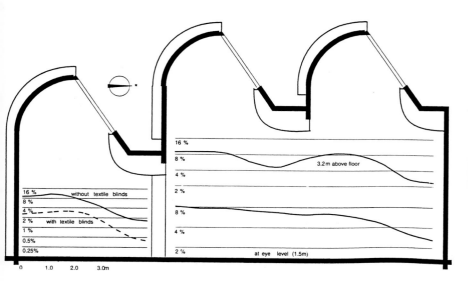

16 %
8 %
4 %
2 %

without textile blinds

16 %
8 %
4 %
2 %
1 %
0.5 %
0.25 %

with textile blinds

3.2m above floor

8 %
4 %
2 %

at eye level (1.5m)

0   1.0   2.0   3.0m

Figure 2.19 The daylight factor distribution in Ludwig museum at Köln, Germany. The use of north-facing monitors and blinds provides daylight control appropriate for the illumination of art objects.

Figure 2.20 Interior view of one of the exhibition areas in the Ludwig museum.

Other specialist building types include sports halls, workshops and exhibition halls. Again it is all too common to find daylight abandoned in favour of more controllable artificial light. In most cases, however, by making full use of the analytical techniques and design tools available, the challenge can be met with rewarding results.

2.2.1 Strategies for existing buildings

The refurbishment and remodelling of an existing building offer a number of opportunities for improving the daylighting of the building (Figure 2.21). However, these opportunities are very dependent on the exact nature of the building and the degree to which it is to be remodelled, and it is difficult to make generalisations.

Considerations of building form, detailed design solutions and their application to refurbishment are dealt with later (Chapter 5.). At a strategic level, the main consideration is the location of the building on the site and its relation to other buildings. This affects the availability of daylight in the same way as already described, except that there may be more constraints in the case of existing buildings. In dense urban environments, it is common to find older buildings severely overshaded by later, higher buildings. Furthermore, the glazed part of the envelope may not currently have an optimum orientation.

The upgrading of windows and the installation of lighting control systems will have little energy benefit in spaces which are so obstructed from the sky that the

Figure 2.21 The refurbished department store Waucquez, in Brussels – now a museum for comic art.

lights have to be on all the time anyway. On the other hand, an increase of glazing area during refurbishment, or the retrofitting of daylight-redirecting devices, or even changes in the reflectance of adjacent surfaces may be able to restore daylight function.

It follows that the potential for 'daylight refurbishment' should be assessed by a visit to the site and, where there is doubt about the influence of other buildings, a modelling exercise should be undertaken.

2.3 Summary

Daylight can be considered to have two components – diffuse light from blue sky and clouds, and direct light from the sun. Direct light is between 5 and 10 times stronger than diffuse light and casts sharp shadows.

Different regions in Europe have different proportions of sunny skies to cloudy skies. For northern and mid Europe diffuse skies predominate, and daylight design is mainly concerned with this condition.

Only southern Europe has predominantly sunny skies. Here, the redistribution of the direct sunbeam by reflection makes a substantial contribution throughout the year and has to be taken into account in daylight design.

The position of the sun in the sky can be predicted accurately for any latitude, time of year and hour of the day. Sunpath diagrams can be used to display this, together with the obstructing effect of buildings and shading devices.

Local obstructions affect both sunlight and diffuse light by blocking parts of the sky. However, in crowded urban sites, reflected light from light-coloured buildings may provide useful daylight. Reflective ground cover, in particular snow, can also make a significant contribution to daylight availability.

The availability of daylight during the period for which the building is occupied is dependent upon latitude, the cloudiness of the regional climate and possibly the degree of local pollution. When considering the potential for daylighting, the function of the building and its occupancy period have to be considered. Daylight availability can be predicted from graphs or tables derived from long-term measurements.

Certain buildings need special consideration. Art galleries and museums, for example, have to control daylight closely to minimise potential damage to exhibits. Other buildings such as sports halls or workshops may have special technical requirements for visual comfort and lighting quality.

Refurbishment and remodelling provide opportunities for improving daylight performance. This may be by upgrading the windows, by altering the glazed area and the position of daylight apertures, or by the inclusion of advanced daylight elements. The impact of existing surrounding buildings must also be taken into account.

References

1. B1. J R Goulding, J O Lewis and T C Steemers, *Energy in Architecture: The European Passive Solar Handbook* (London: CEC/Batsford, 1992), p. 38.

2. N V Baker, A Fanchiotti and K Steemers (eds), *Daylighting in Architecture: A European Reference Book* (London: CEC DG XII/James & James (Science Publishers) Ltd, 1993).

3. P R Tregenza (ed), *European Daylighting Atlas* (CEC/National Observatory of Athens IMPAE, 1997), p. 38 and 77.

4. *Climate of London* (London: Meteorological Office, 1953), p. 124.

5. Tregenza, *European Daylighting Atlas*.

6. Baker et al., *Daylighting in Architecture*.

7. Tregenza, *European Daylighting Atlas*.

8. *Ibid.*

3 Building form and the potential for daylighting

**The NMB Postbank
Head Office,
Amsterdam.**

Building form and the potential for daylighting

Daylight – sunlight and skylight – is admitted into buildings through apertures. For a given facade, daylight of a sufficient level to be useful will penetrate the building up to a certain depth. Thus the dimensions of the building, in both plan and section, have fundamental implications for the degree to which it can be daylit.

The site can pose constraints on the choice of built form, which will influence the possibilities for optimising the daylighting. Clearly the shape and size of the site may influence the shape of the building, particularly in dense urban situations. Another important influence is the obstruction to the sun and sky from terrain and other buildings. Legislation and planning codes may also impose some restrictions on building form.

In spite of the relevance of the daylight climate (in particular the difference between sunny and cloudy climates), from a European perspective it appears that building form is not strongly predetermined by this. The daylighting strategies relating to plan and section, and the use of secondary spaces such as atria and courtyards, find their place in a wide range of European climates. However, climatic differentiation will begin to show at a more detailed design level – relating to the size and orientation of openings, the provision of shading etc, as well as other design considerations such as the provision of natural ventilation.

Figure 3.1 The shallow plan of Wren's Trinity Library at Cambridge (a) is a response to the need for daylight. This contrasts with the deep-plan office block at Canary Wharf, London (b), which, in spite of its fully glazed facade, is artificially lit. (London Docklands Development Corporation)

a

b

3.1 The plan and section

In making the first steps in the development of the overall building form, it is unusual for the architect to see daylighting design as a primary consideration. In many cases, constraints due to the size of the site and the total built floor area requirements of the brief will dictate, or at least limit, the range of options. Where site area is not so constrained, then other planning and organisational issues will normally take precedence over considerations for daylighting.

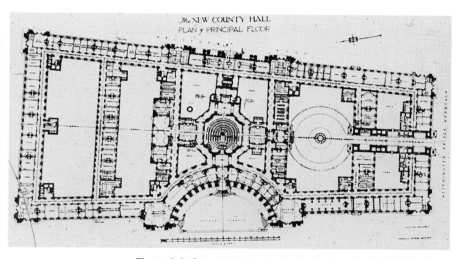

Figure 3.2 County Hall, Westminster, London, by Knott (1908). This large civic building, monolithic from the outside, is in fact a series of shallow plan sections ranged around courtyards.

Figure 3.3 An aerial view of the NMB headquarters at Amsterdam. Every workstation is less than 6 m from a window. This has resulted in a shallow and convoluted plan (NMB Postbank Group).

However, it is as well to realise that early design decisions can have a considerable impact on the potential use of daylight. And it may be that, with increasing realisation of the benefits of daylight from both an energy and an environmental viewpoint, we shall find the daylight criterion moving up the list of priorities in the near future. We might expect to see this occurring first in buildings where daylight is perceived to have a value other than simply for energy saving. Buildings such as schools, hospitals, galleries and museums might be expected to fall into this category.

The principle is simple. Daylight has to penetrate the interior of a building either from the vertical facades (i.e. sidelighting) or from the roof (zenithal or rooflighting). It follows that the key parameter becomes the building depth.

3.1.1 Sidelit spaces

For sidelit multi-storey buildings, the plan depth is the critical parameter. As a rule of thumb, without advanced daylighting techniques a room can be adequately daylit for a depth (distance from the facade) equal to twice the floor to ceiling height (strictly twice the floor to top of window height). For typical floor to ceiling heights of around 3 m this means that buildings up to 12 m deep can potentially be daylit, provided both facades have access to unobstructed daylight.

Historically, this simple rule was well understood, and shallow plans, even in large buildings, were achieved by courtyards and lightwells, as illustrated by the County Hall in Westminster, London (Figure 3.2). Recently, limitation of plan depth has returned as a form generator, with building codes restricting the distance of workstations from a window to 6 m (Germany, Denmark and Netherlands). The NMB bank in Amsterdam (Figure 3.3), by Ton Alperts, is a dramatic illustration of this with its highly convoluted outline.

The daylighting potential of a particular plan can be evaluated to an approximate level of accuracy by taking the ratio of the gross floor area that is fully daylit, i.e. the area within 5 or 6 m of the perimeter, to the total area. Table 3.1 shows a number of plan forms for a 4000 m² four-storey building and indicates the potential daylight performance. This is the basic principle of the LT method (8.2.7), where the availability of daylight is evaluated taking account of the local sky conditions and the occupancy pattern of the particular building.

Table 3.1: Potential daylit fraction for different plan aspect ratios for 4000 m² building with 5 m daylit perimeter zone.

Plan Length to width ratio	Daylit floor area (%)	Short side length (m)
1:1	53	31.6
2:1	57	22.4
3:1	63	18.3
5:1	75	14.1
8:1	90	11.2
10:1	100	10.0

3.1.2 Rooflit spaces

Single-storey buildings and the top floor of multi-storey buildings have no constraint on plan depth since they can be lit through the roof surface. This is often referred to as zenithal daylighting.

Although the glazed openings are distributed over a horizontal surface of the building envelope, the glazed aperture itself does not necessarily have to be horizontal.

Various configurations – sheds, monitors and saw-tooths – employ glazings, which are inclined or vertical, with specific orientations to avoid or encourage solar gain. The inclination may influence which part of the sky provides the light, but it does not alter the fact that the daylight can subsequently be distributed over the entire floor, irrespective of plan depth.

In northern climates, where the brightest part of the overcast sky is immediately overhead, horizontal roof apertures are found more frequently. Note that it is the inclination of the aperture that is of prime importance; the glazing itself may be inclined or curved, but this has less influence on the transmitted light (Figure 3.4).

A good example of horizontal aperture rooflighting is to be seen in the London Stansted Airport Terminal (Figure 3.5). Simple horizontal apertures provide daylight over the whole floor, which measures 196 m x 162 m. Semi-translucent screens reduce sky glare and reflect light onto the ceiling. In southern climates with higher sun angles, more sunshine hours, and more serious consequences from solar gain, horizontal apertures should be avoided.

The design of rooflights is considered in section 4.3.8 and of shading for rooflights in 7.2.

Figure 3.4 Paustian House, Copenhagen. The glazing to rooflights may be inclined or curved. The daylighting function is governed mainly by the inclination and orientation of the aperture rather than the glass surface.

Figure 3.5 Stansted Airport, UK. Simple horizontal apertures with internal shading devices give daylight to the whole floor plate.

3.2 Orientation and obstructions

The term 'orientation' (of a building) usually implies the direction that the main facade or facades face. This is still ambiguous unless we define the main facades as those with the majority of glazing area. In buildings that have a non-square plan, we usually find that it is the longer facades that are glazed, and hence these would be referred to as the main facades. There can remain some confusion, however, since the phrase 'the building is orientated east–west' could mean that the longest axis of the building lies east–west. To avoid confusion we shall refer to the building facing a direction rather than being orientated in that direction.

Irrespective of climate, there are advantages for the main facades of a building to face north and south, rather than east and west. This is because the sun is low in the sky in the east and west, even in the summer, which makes shading difficult, and impossible if a view is to be maintained. On the other hand, north-facing windows receive direct sunlight only in high summer, early in the morning and in the evening, whilst south-facing facades can be easily shaded by small overhangs, due to the high angle of the sun when it is in the southern sky (Figure 3.6).

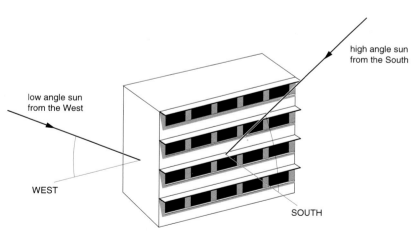

Figure 3.6 An orientation with the main facades of the building facing north and south is preferable. South-facing windows can be easily shaded by overhangs from the high summer sun. East- and west-facing windows are difficult to shade from the low-angle sun.

As well as simplifying the task for providing daylight without excessive solar gain in summer, a south-facing facade does allow the ingress of low-angle sun in winter, when, provided that the visual problem of glare can be tolerated or dealt with internally, the thermal benefit of the solar gain can be realised.

A possible objection to this orientation is that only the south-facing rooms receive sunlight in winter, which, if winter sunshine is seen as a positive quality, disadvantages the north-facing rooms. There is also a case for directing the 'solar' facade to the south-east, so that the solar gains arrive earlier in the day when the building and ambient temperature is lower. Due to the timing of the exposure to the sun, there may also be a health benefit to the occupants (10.2). This increases the utilisation of solar gains and reduces the risk of overheating. It will, how-ever, marginally increase the problem of glare control.

The detailed design of overhangs and other shading devices is dealt with in section 7.2, and sunpath diagrams, which can be used to test the performance of geometric devices, are explained in 2.1, and provided in 11.7.

3.2.1 Orientation and the overcast sky

The standard overcast skies, the CIE sky and the Uniform sky are described in the glossary. These 'design skies' are symmetrical about a vertical direct axis, i.e. they would cause the same illuminance on a vertical surface of any orientation. Real skies are brighter in the south (that is excluding the effect of direct solar radiation) and cause higher illumination in south-facing sidelit rooms.

The difference is quite significant. Hunt has published the orientation factors for the UK (Table 3.2); these show that the relative availability of daylight varies by about ± 20%. The use of these factors, and how they are used to calculate the effect on the demand for artificial lighting, is described in section 8.2. Values for European zones are also given.

Table 3.2: Orientation factors for daylight availability in the UK.

Orientation	Factor
North	0.77
East	1.04
South	1.20
West	1.00

3.2.2 Obstructions

The effect of obstructions has been briefly described in section 2.1.2. Figure 2.17 showed the effect of the height of an obstruction, or strictly speaking the angular elevation of an obstruction, on the energy demand for artificial light. The effect is quite strong: for example, for an obstruction with an elevation of 30°, the lighting energy demand can be increased by 38%.

In placing a building on a site where significant obstructions exist on the boundary, this effect should be given careful consideration. The relevant parameter is the angle of elevation of the obstruction from the window wall, and this will vary for different storeys and, for a given height, the angle of obstruction will vary with distance from the obstruction.

In order to maximise the availability of daylight to as much floor area as possible, the objective is to provide a sufficiently large view of the sky from as many windows as possible. The view of the sky can be achieved either above the obstruction, or at the side or between obstructions, as illustrated in Figure 3.7.

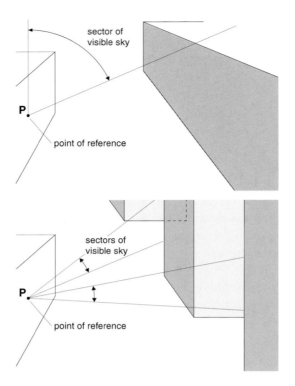

Figure 3.7 The principle of daylight access. Daylight can be provided by sufficient views of the sky either above adjacent buildings, or between them, or both.

3.3 Atria, lightwells, courtyards and galleria

The benefits of adopting a shallow plan when designing a sidelit building have already been discussed, together with the fact that single-storey plans of any depth can be provided with rooflights. In this section we now consider the degree to which the advantages of providing daylight from above can be extended to deep-plan multi-storey buildings.

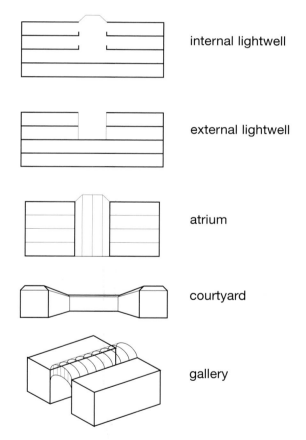

Figure 3.8 Definition of internal and external lightwells, atrium, courtyard and gallery.

internal lightwell

external lightwell

atrium

courtyard

gallery

A range of design solutions have evolved and become part of the common architectural vocabulary – courtyards, lightwells, galleria and atria. However, the terms are often loosely applied, and it is useful to define an unambiguous nomenclature to be used here (Figure 3.8).

These solutions are to some extent influenced by climate. Clearly the open courtyard exposes a large surface area of the building to the relatively unmodified outdoor climate, and in cold climates this would lead to heat loss. Furthermore, the open space would be unusable for most of the year. Thus the courtyard solution suits the milder climates of southern Europe, and indeed this is where this form is adopted most by vernacular architecture. On the other hand, an enclosed atrium or galleria can provide useful temperature increments in the winter, as well as protection from rain and wind, much appreciated in the colder climates. But they require shading in summer to prevent overheating, especially in southern climates.

If we look at existing buildings, the incidence of galleria and atria does seem to be greater in the north, and the incidence of courtyards greater in the south. However, the courtyard form is also found all over Europe in many older buildings (up to 1940), indicating their response to the need for daylight and natural ventilation of the parent building, rather than their climate-modifying function (Figure 3.9). The thermal performance of these elements is described in more detail in *Energy in Architecture* [1].

Figure 3.9 Neville's Court at Trinity College, Cambridge, UK. The open courtyard form is found in northern Europe in spite of the thermal disadvantage of the increased surface area.

In order to regard them as daylighting features, we must consider their function in providing daylight to the adjacent spaces. However, particularly in the case of atria and galleria that are themselves normally semi-occupied, the daylighting in the atrium or gallery itself must also be adequate. Unfortunately, some modern atrium designs have such poor daylighting characteristics that, in order for planting and vegetation to thrive, artificial lighting has to be provided even in the daytime (Figure 3.10). Clearly, if the daylighting in the atrium or galleria is inadequate, then it is very unlikely to make any contribution to the surrounding spaces.

Figure 3.10 Permanent artificial lighting used to ensure survival of indoor planting in poorly daylit atrium.

3.3.1 Sky component and reflected light

We shall now describe the technical principles of the provision of daylight from these elements. For convenience we refer mainly to the atrium, but make reference to the courtyard, galleria and lightwell where there are significant differences. We shall deal first with daylighting from the diffuse overcast sky, then describe the case for predominantly sunny skies.

For sidelit rooms without special daylighting devices (Chapter 9), it is possible to provide good levels of daylight (DF >2%) for a distance from the facade of about twice the floor to ceiling height. This is assuming that about 40–50% of the total wall area of the facade is glazed and that the sky is unobstructed above 10°. For a facade facing an atrium (except in the case of the 'lean-to' atrium) the other separating walls of the atrium will heavily obstruct the sky. This is particularly true for a deep internal atrium as shown in Figure 3.8.

Owing to the high level of obstruction usually encountered in a top-lit atrium, the view of the sky from the workplane of surrounding rooms is very limited, and is further reduced in average luminance by the absorption of the glass roof and the obstruction of the structure that supports it. In the absence of reflected light, this will have the effect of severely reducing the depth of the daylit zone, as illustrated in Figure 3.11.

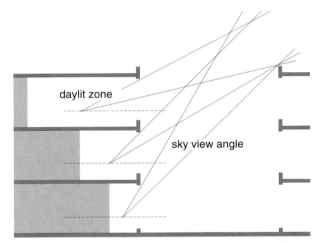

Figure 3.11 Influence of the opposite walls of the atrium on depth of the daylit zone in the adjacent rooms.

From the geometry, it is clear that the daylit zone has to become progressively more shallow, the lower the storey, in order to maintain a given view angle of the sky. Good daylighting design may respond to this by increasing the size of the windows (4.3).

However, the opaque surfaces of the facades and the floor of the atrium get a relatively good view of the glazed roof and, if they are light in colour, can become useful secondary sources of light themselves as well as the sky. The reduction of window area in the upper stories facing the atrium, owing to their less obstructed view of the sky, leaves a larger area of opaque wall to provide reflecting surfaces to conduct light to the lower levels of the atrium. The three critical components – light from the sky, light reflected from the walls and light reflected from the floor of the atrium – are shown in Figure 3.12.

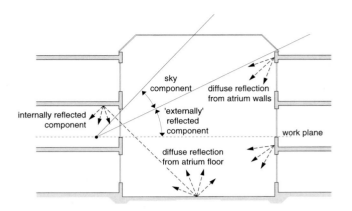

Figure 3.12 Useful daylighting in the rooms adjacent to an atrium consists of light from the sky, light reflected from the walls of the atrium, and light reflected from the floor.

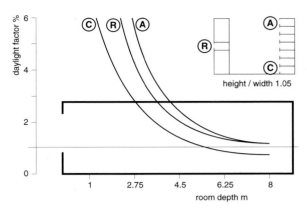

Figure 3.13 Daylight factor in adjacent rooms as a function of atrium parameters. Values for atrium-facing rooms A and C compared with unobstructed reference room R. (Willibold-Lohr)[2]

To summarise:

- The relative magnitude of these light components is dependent on the geometry, which is different for the rooms at the top of the atrium than for those at the bottom. This leads to the conclusion that glazing areas should increase for the lower rooms facing the atrium.

- The average reflectance of the atrium walls is reduced by the glazing (glazing has a reflectance of only about 15%). The more glass, the lower the average reflectance (except where the opaque wall is of very low reflectance).

- Light coming from the atrium below the horizontal will fall on the workplane only via reflection from internal walls, or more likely the ceiling. Thus the reflectance of these surfaces also becomes critical:

The following parameters determine the daylight factor in the rooms adjacent to the atrium:

- atrium height to width ratio (above the room being considered)

- transmittivity of atrium external glazing

- average reflectance of atrium walls

- average reflectance of atrium floor

- glazing ratio of atrium walls

- reflectance of room ceiling and walls

From various model studies and data from the case study buildings, a design tool responding to these parameters has been developed by Willibold-Lohr. A sample graph is

shown in Figure 3.13. Note that the daylighting in room A is slightly better then the reference case. This is due to useful reflections from the upper atrium wall, which in this study was assumed to be white.

3.3.2   Lightwells

The lightwell has been defined in Figure 3.8 and can be considered as a top-lit atrium without a wall separating the atrium from the surrounding spaces. This has significant implications for ventilation and thermal performance of the space, but this does not concern us here. The impact on the daylight environment is primarily due to the absence of glass and the fact that there may be even less opaque surface surrounding the lightwell than in the case of the atrium. Glass has a reflectance of about 15% increasing rapidly at oblique angles (6.1.2), the latter leading to some useful redirection of light from high up on the atrium walls. In a lightwell, the open spaces between the floors have, in effect, virtually zero reflectance, which means that reflection from the floor and opaque balustrades is of increasing importance (Figure 3.14).

In daylighting function the external variant of the lightwell behaves like a courtyard (see below) with a very large height to width ratio.

3.3.3   Courtyards

Courtyards have no roofs and thus do not suffer from the reduction of sky luminance due to obstruction and absorption of the glass. This could easily lead to a doubling of the light available compared with a glazed atrium. On the other hand, there is usually no roof-level structure on which to support shading devices, and this

Figure 3.14 The lightwell at the Anglia Polytechnic University Learning Resource centre, Chelmsford, UK. Unlike an atrium, the separating walls are unglazed. Reflective surfaces redirect light through the unglazed openings.

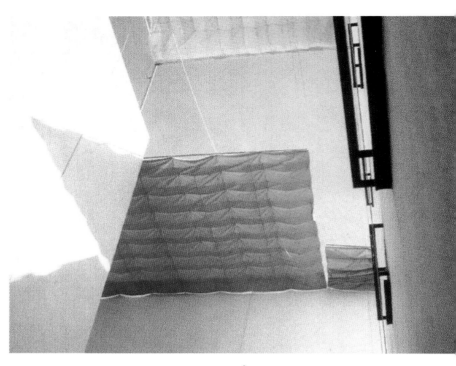

Figure 3.15 A shaded courtyard in Seville, Spain.[3]

could result in excessive glare under sunlight conditions if highly reflective finishes are used. In some cases, open courtyards are shaded, as illustrated in Figure 3.15. External finishes tend to be darker than internal finishes, particularly in polluted urban environments.

Since courtyards are open to the weather, it is common to find the adjacent buildings with cloisters or verandas, offering some protection from both sun and rain. These will shade the glazing of the adjacent rooms from direct sun, but they will also significantly reduce the sky view from these rooms. This means that if reasonable daylight levels are to be achieved in the rooms, ground-reflected light will have to make a major contribution, particularly since the surrounding walls will be in shadow and therefore will not act as diffuse sources (Figure 3.16).

In sunny climates, and where the courtyard ground finish is light in colour (e.g. gravel, paving, etc), good daylighting can be achieved with the cloister configuration. In more northern, cloudy climates, lower illuminance from the diffuse sky, together with ground finishes that tend to be darker (often including grass), daylighting in the adjacent rooms may become difficult. One possible solution is the projecting cloister with clerestory windows above, lighting the adjacent rooms.

Figure 3.16 Rooms adjacent to cloistered courtyards rely heavily on ground-reflected light, Venice.

a

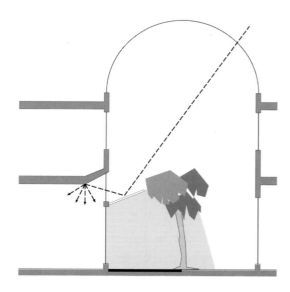

Figure 3.18 The control of the luminance of daylit surfaces in the galleria to optimise viewing into adjacent display windows.

b

Figure 3.17 The St Hubert Galleries, Brussels. The illuminance in the galleria (a) is at least 10 times higher than the display space (b), usually resulting in the use of display lighting.

3.3.4 Galleria

Many of the issues that apply to atria also apply to galleria, such as the effect of the glazed covering in reducing illumination, and the dependence on ground-reflected light and on light reflected from opposite walls. However, the frequent use of galleria as shopping or exhibition areas makes a special demand on the lighting environment. This is the function of the gallery as a covered street from which the interiors of the adjacent display rooms are viewed through windows.

In the classic section (Figure 3.17), the luminance of surfaces inside the adjacent room and viewed through the glazing will be typically 10 times lower than the surfaces in the galleria itself. This will be exacerbated by employing light-coloured surfaces in the gallery to maximise reflected light. The result of this is that goods and exhibits viewed through the windows will have very low visual impact and, inevitably, artificial lighting will be used to augment the ambient daylight. Since this will often be tungsten display lighting, it will result in a large increase in energy consumption and often unwanted heat gains. In this almost universal case, it can be seen that the interpretation of galleria as 'daylighting solutions' is optimistic to say the least.

However, by manipulating the relative luminances of the display areas and the street area, more use of daylight could be made. Some possible configurations, similar in principle to the clerestory lighting suggested for the courtyard, are suggested in Figure 3.18.

3.4 Summary

Daylighting from the vertical facade can usefully penetrate for a distance of twice the floor to ceiling height. This means that, typically, buildings of up to 12 m total depth can be daylit, provided that the facades receive unobstructed light from the sky and that the interior furnishings or partitions do not cause major obstruction (Chapter 4).

This distance of useful daylight penetration may be increased if advanced daylighting elements are used (Chapter 9).

For single-storey buildings, or the top floor of multi-storey buildings, the whole floor area can receive daylight from the roof. Roof lights may have the aperture in a horizontal plane, or be constructed with the aperture at any inclination and orientation in order to control sunlight penetration. Roof lighting geometry may become an integral part of the roof structure.

To a limited extent, daylight can be brought into deep-plan multi-storey buildings by means of atria and lightwells. Their effectiveness in providing light to adjacent spaces is very dependent upon the width to height ratio of the atrium or light well, and the reflectances of the opaque surfaces within the atrium or lightwell.

Courtyards provide access to daylight by avoiding deep plans in buildings with large floor area, but without the major climatic modification of the atrium. Courtyards with surrounding cloisters rely on ground-reflected light to penetrate the adjacent rooms.

Building form is not highly sensitive to climate in the way it responds to the need for daylighting. Response tends to be more at the level of detailed design of orientation, opening size, shading and re-directing devices.

One exception to the above is that atria with large areas of horizontal roof glazing should be avoided in warmer European climates.

References

1. J R Goulding, J O Lewis and T C Steemers, *Energy in Architecture: The European Passive Solar Handbook* (London: CEC/Batsford, 1992).

2. N Baker, A Fanchiotti, K Steemers (eds) *Daylighting in Architecture* (London: James & James (Science Publishers) Ltd, 1993).

3. Energy Conscious Tradition, from a poster by the EU.

4 The daylight design of spaces

Lady writing a letter, with her maid (Vermeer): a master-piece of daylight rendering.

The daylight design of spaces

The internal planning and circulation of a building will affect (or be constrained by) the built form (3.0) The nature of this interrelationship between planning and form is to a certain extent determined by the quantitative daylighting needs of each space, and by the qualitative intentions of the design team. These two aspects of lighting criteria can be strongly related, particularly in terms of dealing with the transition between spaces, and the setting up of a luminous hierarchy between the outside and the interior.

4.1 Transition and differentiation

4.1.1 Temple or church: heightening of sensibilities

This is perhaps most clearly evident in historical architectural examples such as the Pantheon (Figure 4.1).

Here, the transition from the exterior to interior space is modulated by the peristyle. The narrow sunlit streets give way to a colonnaded portico, which shades the entrance and allows the eyes to adapt before one enters the expansive top-lit dome. The phenomenon of adaptation is discussed in more detail in sections 6.2.4 and 10.1.2.

It is as if one enters from between the tree trunks of a wood into a clearing lit by a shaft of light. The qualitative manipulations, concerned with the heightening of sensibilities as the visitor approaches the temple, can be expressed quantitatively by the change of daylight conditions from the relatively bright street, via the colonnade with shafts of sidelight, into the subdued toplit dome. Not only does the quantity of daylight change in response to the architectural intentions, but also the predominant direction of light. The quantitative criterion for the main interior is to provide low-level light, suit-

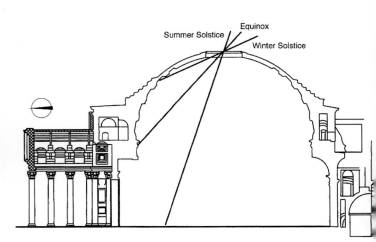

Figure 4.1 Interior view (a) and section (b) of the Pantheon, Rome. The simple geometry of the single oculus causes the dramatic sunlighting of the interior.

Summer Solstice

Equinox

Winter Solstice

a

b

able for contemplation and worship, where the source is separate from normal sidelighting and not related to the city, but to the heavens.

4.1.2 School or office: functional response

At a more prosaic level of, for example, the workplace environment, this issue of the interrelationship between spaces and the lighting criteria will have implications for internal planning and room layout. The transition from corridor to office, or corridor to classroom, will have an important impact on the visual perception of the space entered. This initial impact may be that a room appears relatively too bright or glaring (10.1) or conversely too dark or dull.

These impressions can affect the use of artificial light and thus energy consumption. It may well be that illuminance criteria are met and that as the occupant adapts the conditions would become acceptable. However, if the first impression on entering the space is that it is gloomy, there is a high probability that the lights will be switched on, and left on, even if after adaptation the illuminance is sufficient.

A good example of such a relationship is provided by the Collège La Vanoise. Here, because the classrooms are generously daylit, the circulation space too is well lit (Figure 4.2). The rooflit circulation or atrium space is bright and sunny, and the classrooms are carefully designed to be light but free from glare.

A smooth transition between spaces is further helped by allowing daylight to enter from the atrium into the back of the classroom through high-level glazing (Figure 4.3). The result is that the functional requirement of lighting classrooms is met (Figure 4.4), i.e. adequate daylight and an even distribution, the circulation area has a more dynamic lighting quality (sunlight is allowed and less control is provided), and the transition is gradual. The provision of lighting at the back of the room from the atrium has allowed classrooms to be double-banked (that is back-to-back). This strategy has had a major impact on both the building form and internal planning.

In many building types the functional requirements with respect to the uses will vary over a wide range. In such cases the relationship between such uses will need particular attention. An example of this is the library building.

Figure 4.2 View of the well-lit circulation space of the Collège La Vanoise, Modane, France.

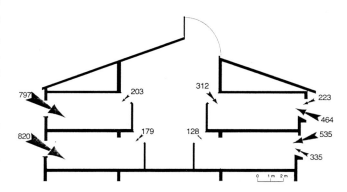

Figure 4.3 Section of the Collège La Vanoise, showing the backs of the classrooms being lit by secondary daylight from the atrium circulation space. The numbers are the light fluxes in lumens, for a total of 1000 lumens of incoming flux to each classroom.

Figure 4.4 Lighting quality in a classroom of the Collège La Vanoise. Note the secondary light from the sloping glazing on the left and the light from the highly reflective room surfaces.

4.1.3 Library: reading and protecting

The dual role of libraries is firstly to store and protect books and secondly to allow access to the books and allow them to be read. A third function of libraries may be to allow the use of computers. Each of these roles requires different lighting conditions: the book stacks need adequate but controlled light on the vertical plane; reading areas require good light on the horizontal plane, possibly associated with views out; and computer use requires relatively low-level, glare-free light.

This differentiation has, both through history and in modern examples, been dealt with in a number of different ways. Historically, the library originated primarily as a reading room, containing relatively few books but many reading benches. Light was provided by daylight, through tall windows. As the number of books available to the scholar increased, the library became as much a building for storing books as a reading environment. Shelves, typically perpendicular to the wall, defined reading carrels open to the main space (e.g. Trinity Library, Figure 4.5).

Increasingly books, and particularly rare book collections, are demanding greater care in environmental terms. New information and retrieval systems, via computers, introduce new lighting criteria. Even in small libraries these distinctions are evident, as can be seen in the Darwin College Library (Figure 4.6).

The corridor on the north side contains most of the books, which receive borrowed light from the south and

Figure 4.6 View of the Darwin College Library, Cambridge, interior from the upper reading area.

from a high-level, north-facing strip of windows. Daylight levels in this area are not excessive, and much reaches the bookstacks indirectly via reflections from adjacent walls. The main reading desks are located on an upper level on the south side with large areas of glazing. Daylight levels are high (Figure 4.7), and the glazing offers views across the River Cam to a green space. Indeed, the unshaded south glazing often leads to glare problems here when direct sunlight falls on the reading matter. However, in user surveys, it seems that this is tolerated largely due to the enjoyment of the magnificent view.

The computer rooms are on a lower level with relatively small areas of glazing and thus low daylight levels, as is

Figure 4.5 Internal view of Trinity Library, Cambridge, showing the effective distribution of light from the upper part of the room with its highly glazed reflective surfaces.

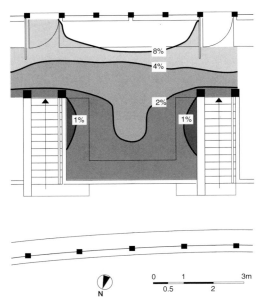

Figure 4.7 The daylight factor map shows the strong reduction in lighting level in the area used for VDUs compared with the reading area close to the south facade.

more appropriate for VDU use. It is through such spatial manipulation of the planning and room layout that the different daylighting requirements are met and clearly expressed architecturally.

Figure 4.8 Carefully controlled toplight at the Neue Staatsgalerie, Stuttgart, Germany.

Numerous examples can be cited that express the need for distinct lighting conditions, in terms of building form, organisation of plan and room layout. Museum buildings, with the particular need to display and preserve artwork, typically adopt carefully controlled toplight for exhibiting paintings (Figure 4.8), and strong sidelight to reveal the three-dimensional qualities of sculpture, for example.

For most buildings the primary concern is to exceed a minimum daylight illuminance. In museums that house objects that can be damaged by light, however, the primary concern is to limit the illuminance to a maximum permitted value, or strictly a maximum light exposure – i.e. the product of illuminance and time.

At a more prosaic level, factory buildings often combine large-span toplit halls for glare-free industrial use with simple sidelit administrative offices (Figure 4.9).

Figure 4.9 Rooflights integrated with the structure and servicing span a large industrial hall in Milton Keynes. (Arch: ECD Partnership)

4.2 The components of daylight

Light is of critical importance in experiencing space. The same room can be made to give very different spatial impressions by the simple expedient of changing the size and location of its openings. These qualitative impressions are measurable in terms of daylight levels and distribution on surfaces enclosing or within a space. It is thus initially easier to discuss the quantifiable characteristics of daylight in a space, and subsequently consider the aesthetic implications.

There are clearly an infinite number of ways in which an opening can be sized and positioned with respect to a space. In order to discuss the subject it is thus appropriate to consider a limited number of generic conditions (Figure 4.10).

4.2.1 Light from the sky

Before examining these different configurations in detail, it is useful to explore the process by which light from the sky reaches a particular point in the room.

Figure 4.11 shows that we can consider the light falling on a point (in this case on a desk or table) in the room as being composed of three distinct components. Light that comes directly from the sky is called the sky component (SC), light that comes from external surfaces such as buildings is called the externally reflected component (ERC), and light that is reflected from internal surfaces is called the internally reflected component (IRC).

The sky component normally refers to the diffuse sky: i.e. it is not used to describe direct sunlight. Obviously, it depends upon there being a view of the sky from the point in the room being considered. It was this fact that forms the basis of a rule by Vitruvius in Book VI of *De Architectura*: "On the side from which the light should be obtained, let a line be stretched from the top of the wall that seems to obstruct the light to the point at which it is to be introduced, and if a considerable space of open sky can be seen when one looks above the line, there will be no obstruction to light in that situation" (Figure 4.12), which is probably the earliest example of a daylighting design rule. Although it may have lost something in the translation, we can see what he is getting at.

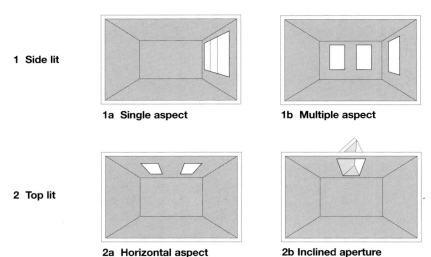

1 Side lit

1a Single aspect   1b Multiple aspect

2 Top lit

2a Horizontal aspect   2b Inclined aperture

3 Secondary

3 Secondary light from an
adjacent space sometimes
referred to as 'borrowed light'

Figure 4.10 Generic configurations for daylighting rooms.

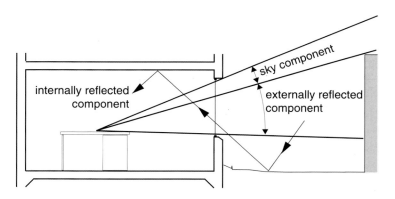

Figure 4.11 The daylight at a point in a room comes from the sky (sky component), reflection from outdoor surfaces (externally reflected component), and reflection from the room surfaces (internally reflected component).

It can easily be appreciated that the view of the sky gets larger as the point considered approaches the window. Thus it is mainly the sky component that leads to the strong variation of light intensity in a sidelit room.

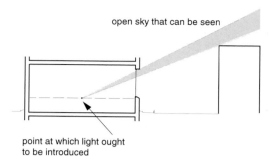

open sky that can be seen

point at which light ought
to be introduced

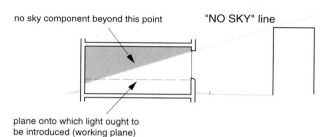

no sky component beyond this point   "NO SKY" line

plane onto which light ought to
be introduced (working plane)

Figure 4.12 Vitruvius' geometric definition of when an interior point is considered to be 'well-lighted' resembles today's rule-of-thumb (the 'no sky' line) beyond which it is probable that insufficient daylight will be available.

4.2.2 Externally reflected light

The externally reflected component is particularly relevant in dense urban situations, where, owing to the closeness of buildings, a view of the sky may be limited or even completely absent for all but positions very close to the window (Figure 4.13). Note that the ERC will tend to come from a low angle, close to horizontal. This will penetrate deeper into the space than the sky component, but owing to the absorption of light by the external obstruction it will generally be much weaker.

4.2.3 Internally reflected light

The internally reflected component (IRC) always reaches the point being considered after having been reflected from a surface within the room. However, these surfaces may have been illuminated either by direct sky light or by light reflected from external surfaces including the ground. It is obvious from Figure 4.11 that any light that is reflected from below the horizontal must be reflected a second time on the ceiling or upper walls of the room, in order to illuminate the horizontal (upward-facing) plane, and will thus end up as the internally reflected component.

An important characteristic of the IRC is that it is fairly uniform all over the room. Figure 4.14 shows how the

Figure 4.13 Further back in a sidelit room, most of the light may come as reflections from neighbouring buildings. (UAP Insurance building, Lausanne, Switzerland)

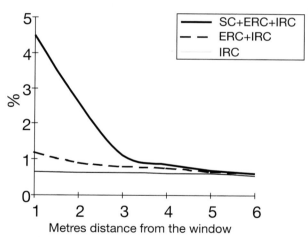

Figure 4.14 Relative contributions of sky component (SC), externally reflected component (ERC) and internally reflected component (IRC) for a typical room with an external obstruction.

relative contributions of the components vary with the distance from the window. At the back of the room it is often the case that there is no SC and sometimes, in an L-shaped room for instance, no ERC. Clearly, the IRC then becomes very important.

It is interesting to note that, well into the 20th century, daylighting was designed by considering the SC only. This was partly due to the fact that interior surfaces tended to be very dark, often as a result of soot emissions from oil and gas lighting, or to mask their effect (Figure 4.15). Modern daylighting design, however, is concerned much more with maximising the use of internally reflected light, often in conjunction with special devices, which re-direct the light onto the ceiling and upper walls (Chapter 9).

Internally reflected light is dealt with in more quantitative detail later (11.6).

4.2.4 The daylight factor

In what units are the components of daylight measured? Clearly daylight illumination could be measured in lux. However, since the outdoor illuminance from the diffuse sky is varying over such a wide range (from about 2000 lux on a really gloomy day to 50 000 lux for a summer day of hazy sunshine), this would always have to be specified as well. Furthermore, owing to the adaptive properties of human vision (10.1.2), except at minimum thresholds, the absolute illuminance is less important than the relative illuminance. Thus it is more convenient and meaningful to specify the daylight illuminance in a room as a ratio (Figure 4.16). This ratio, usually expressed as a percentage, is called the daylight factor (DF).

The DF is simply the sum of the three components described above:

$$DF = SC + ERC + IRC$$

To measure the DF in an existing building the illuminance is measured inside, at the point of interest, and simultaneously (or within as short a time as possible) outside under the unobstructed diffuse sky. The ratio of the two values is then calculated. Typically DFs are quite small, daylight being useful for DFs as low as 1%. A room with an average DF of 5% would be regarded as brightly daylit.

Measurements such as these were carried out in all the case study buildings. Because of the large variation of DF within a room it is informative to plot contour maps as in Figure 4.17. The interpretation of these maps is discussed later.

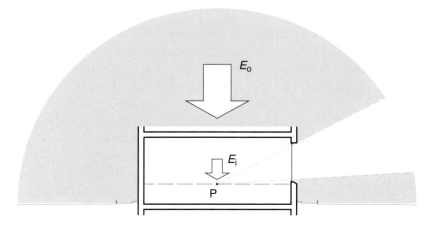

Figure 4.16 Daylight factor is defined by DF = E_i / E_o x 100%, where E_i is the internal illuminance on the horizontal plane, and E_o is the external unobstructed illuminance on the horizontal plane.

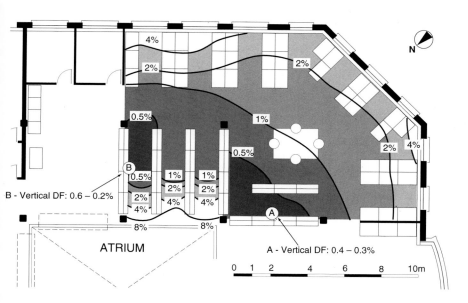

Figure 4.17 Daylight factor (DF) contour map of reading and bookstack area of the library at Anglia Polytechnic University, UK.

Table 4.1: Selected daylight factors

Building type	Location	Average daylight factor (%)	Minimum daylight factor (%)
Airport buildings	Reception areas	2	0.6
	Circulation areas	2	0.6
General building areas	Reception areas	2	0.6
	General offices	3	1.0
	Drawing offices	5	2.5
Libraries	Stacks	3	1.0
	Reading areas	6	1.5
Museums and art galleries	General	5	1
Schools	Classrooms	5	2
	Assembly halls	1	0.3
Domestic	Multi-purpose rooms	1.5	0.5
	Kitchens	2	0.6

For a more detailed version of this table please see CIBSE: *Window Design Applications Manual*

There are several ways to predict the DF for a room from the specification of the geometry, the surface reflectances and the surroundings. Some of these will be described in 4.4. Most methods involve the calculation of the three components and their summation.

4.2.5 Typical values of daylight factor

Table 4.1 shows some recommended values for minimum and average daylight factor for various functions. These have been derived for overcast skies, predominantly in N Europe. In S Europe, where skies are brighter and where direct sunshine make a significant contribution to useful daylight, we would expect smaller values.

To those unfamiliar with the quantification of daylight, it may come as a surprise to see such low values. We are not normally aware of such differences between indoor and outdoor illuminance owing to the adaptive ability of our eyes. However, the difference does become apparent when moving suddenly from the bright outdoor space to an interior.

It is because outdoor illuminance is so high that even very small DFs are sufficient. For example, a DF of 1% on a bright overcast day with an illuminance of 12 000 lux (12 klux) gives an indoor illuminance of 120 lux, which is quite adequate for a normally sighted person to read by.

In order to assess the target DF, in terms of what it means over time for the function of the space, we can apply it

to the daylight availability curves (12.2). These are horizontal illuminance data presented in a cumulative frequency form – they indicate the fraction of a specified time period (usually the building occupancy period or else sunrise to sunset) for which a specified outdoor illuminance is exceeded (Figure 4.18).

For example, an office has a minimum DF of 2%. If we consider a minimum illumination of 200 lux is required, then this will be achieved when the outdoor illuminance is 200/0.02, that is 10 klux. By referring to the availability curve, we can see that this is exceeded for 66% of the office working hours over the year. Bearing in mind that

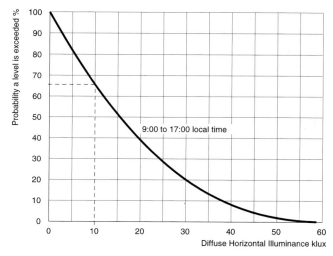

Figure 4.18 Annual daylight availability (diffuse only) for London showing the threshold illuminance of 10 klux (to give 200 lux in the office with a 2% DF) is exceeded for 66% of office working hours.

Table 4.2: Daylight factors measured in selected Daylight Europe case study buildings.

Building	Average daylight factor (%)	Minimum daylight factor (%)
Stansted Airport, UK	3	2.5
Agricultural Bank (south-facing room), Greece	0.6	0.1
Statoil Research Centre (office), Norway	2	0.3
Queens Building (second floor), UK	1.4	0.5
Kew Palm House, UK	18	12
Darwin Library, UK	6	1
Trinity Library, UK	18	0.5
Soane Museum (breakfast room), UK	2	0.5
Dragvold University Centre (second floor office), Norway	2	0.5
Learning Resource Centre (reading areas), UK	2	0.5
Collège La Vanoise (north facing classroom on lower floor level), France	3	0.8
Hawkes' house (living room), UK	5	1

some office hours in winter will occur after sunset, this could be regarded as a moderately well daylit space. This analysis is dealt with in more detail in 8.2.

As part of the Daylight Europe study, DF measurements were made in 12 of the case study buildings. These are presented in Table 4.2. It is difficult to make a comparison with such a small sample, but it is worth pointing out the low value for the south-facing room in Greece and the predominantly higher values in the more northern offices. The very high value in the Kew Palm House is because of its horticultural function, whilst the unusually high value in the Hawkes House is to serve the objective of solar collection. It is also interesting to note that many of these are lower than values recommended in Table 4.1. However, it has been found that the range of DF within a room is often more important than absolute values.

4.3 The configuration of windows

The position, shape and size of windows in a room have a strong influence on the distribution and level of daylight, and hence the usefulness of the daylight. Although we have identified the variation of daylight in a space as a positive quality, in most cases the objective of good window design is to avoid strong variation of illuminance. This objective can be quantified as described later in 4.3.5.

In the following sections, most of the discussion applies to overcast skies, and thus little reference is made to orientation. For mid and N Europe, daylight design is usually based upon this condition. However, that is not to imply that orientation is unimportant in these latitudes; consideration has to be given to the control of direct sunlight by shading – this is dealt with in detail in 7.2.

In climates where the clear, sunny sky is most typical, the design of openings becomes complex. The reason is that, unlike the overcast sky, the clear sky is much more changeable and dynamic with respect to time of day and year. The influence of direct sunlight is of critical importance, contributing to the interior natural light level whilst increasing solar thermal gains and possible glare problems. This is dealt with in greater detail in Chapter 10.

4.3.1 Sidelighting and room layout

The most common way to introduce daylight into a space is via vertical openings in walls to give sidelighting. A critical issue that arises in such spaces is that the daylight distribution is not uniform, falling off rapidly as one moves away from the opening. Figure 4.19 shows the section of a sidelit room, with the daylight factor for the workplane superimposed. Note the rapid increase in DF close to the window as the view from the workplane takes in more sky.

The inherent variation of daylight illuminance in sidelit spaces suggests a corresponding variation of use. It is clear that in a sidelit space the tasks that require greater visual acuity, such as sewing or reading, should be located closer to the opening. Conversely, uses that need lower light levels, such as circulation, storage of books or VDU use, can be located further from the facade. A

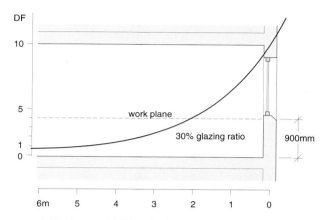

Figure 4.19: The rapid fall in DF with distance from window wall.

similar strategy should be adopted with respect to the location and control of artificial lighting (8.1).

Cellular offices are a case in point where the daylight distribution in sidelit spaces can be exploited (Figure 4.20). The desk, the focus of most activities in an office, is typically directly adjacent to the opening to enjoy the best daylight. If computers are used then these should be placed on the desk away from and perpendicular to the source of daylight. This is to ensure that glare (10.1.6) is minimised and VDU use is comfortable. As most VDUs are themselves luminous, but are of limited maximum brightness, there is an absolute recommended maximum illuminance level of 150 to 250 lux. If and when higher levels exist, they should be reduced by shading, as described in 7.2. Towards the back of the room it may be

Figure 4.20 View of a cellular office at the Statoil Research Centre, Trondheim, Norway.

appropriate to locate storage, which requires lower lighting levels, typically a minimum of 150 lux.

However, the problem is not just that of horizontal illuminance. The luminance of surfaces, particularly vertical surfaces, is also a key factor in functional daylight design. This is dealt with briefly in the next section, and in more detail in 10.1.

4.3.2 Daylight quality and visual comfort

From the discussion above, it becomes apparent that daylighting is not just about quantity – i.e. the more light the better. The concept of quality with respect to daylight is primarily concerned with daylight distribution, as measured at the point of interest, whether this be on a horizontal workplane, or in the case of, for example, an art gallery, a vertical plane. It is also concerned with the distribution of the brightness of surfaces – not just those of the task or object of interest, but also the surrounding surfaces, the views of which contribute to a person's overall perception of the space and satisfaction with it.

The brightness of surfaces is a product of the illuminance and the reflectance of the surface, which introduces a new level of complexity. Shiny surfaces introduce specular reflections - i.e. like those of a mirror, causing both discomfort and actual impairment of vision. A further complication is that the windows themselves form a significant proportion of the room surface area, which may become glare sources if they are frequently in the field of view.

Since it is also the illumination of surfaces that is the main way by which the architecture itself is revealed, and hence the aesthetic qualities, we are clearly dealing with a complex problem.

In the following sections, however, concern is restricted to the distribution of DF on the workplane, the simplest quantitative parameter. Quality and visual comfort is dealt with in much more detail in 10.1.

4.3.3 The location of windows

Although the typical single-sided room has a rapid reduction of daylight from the opening to the back of the room, the distribution will be affected by window proportion and location in the window wall. For a given area of opening, the distribution will generally improve the higher the glazed area is located in the window wall. This is illustrated in Figure 4.21, which is generated by a simulation using the computer program SUPERLITE. Table 4.3 shows the minimum, maximum and mean DF values for the three cases.

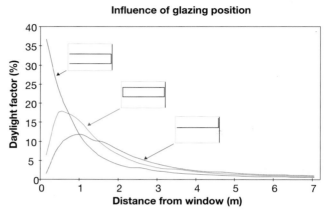

Figure 4.21 Comparison of distribution of high and low openings (SUPERLITE simulation).

Table 4.3: DF values for a room with three different window positions.

Window position	Minimum DF	Maximum DF	Ratio	Mean
Low	0.5	37	74	4.9
Medium	0.6	17	28	4.4
High	0.75	12	16	3.9

The test room is 7.2 m deep, and the figures above show that this is too deep for adequate daylight at the back. Only for the highest window position does the max/min ratio begin to approach an acceptable value of about 10. However, it is clearly apparent that the distribution improves with the higher glazing positions. Note that the mean value is only weakly affected by the window position, but in all cases is adequate.

Thus in rooms where the uses are intended to be similar throughout the plan depth it is appropriate to locate openings higher to provide a more even distribution of daylight.

Similarly, a deeper plan will benefit from both greater areas of openings, but more importantly from high openings. Thus higher ceilings enable deeper daylight penetration. As a rule of thumb, the depth of useful levels of daylight penetration is approximately twice the distance from the floor to the top of the window opening (Figure 4.22), unless special daylighting devices are used (Chapter 9). Since the highest point for the window is up to the ceiling plane, this rule is often quoted as 'the potential daylit zone depth is twice the floor to ceiling height'. Figure 4.23 shows the application of this princi-

Figure 4.22 A simple rule that the useful daylit depth from the window wall is a maximum of twice the height from the floor to the top of the window.

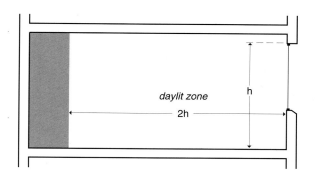

Figure 4.23 Domino Haus, Reutlinger, Germany. Clerestory lighting developing the full potential of the daylit zone depth of twice the floor to ceiling height.

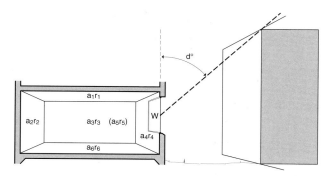

Figure 4.24 The calculation of the average daylight factor by the simplified Lynes method, as described in the text.

ple in the form of clerestory windows extending to the full wall height, providing good daylighting to the back of the room. Adjustable shades control the illuminance close to the windows.

A more precise treatment is suggested by Lynes.[2] He defines a critical ratio:

K = ave DF front half of room / ave DF back half of room

and states that if K is greater than 3, then the back half of the room will appear unacceptably gloomy. He goes on to show that for an average reflectance of 0.5, this criterion is just met in sidelit rooms for a depth of up to twice the floor to ceiling height.

Note that this ratio is different from the DF_{max}/DF_{min} ratio, which for a sidelit room should not be greater than about 10. However, the problem with this ratio is that DF_{max} is difficult to define meaningfully, since its value rises very steeply close to the plane of the window. Because of this the Lynes criterion is recommended.

4.3.4   The size of windows

The appropriate size of openings should relate to the required internal illuminance (related to the activity) and the typical sky conditions. Thus for office use in southern

climates, where the clear sky predominates, it is likely that the opening sizes should be smaller than for northern overcast skies. The variation in daylight availability and sky type has already been discussed (2.1).

The calculation of daylight factor for overcast skies, and its distribution over the room, using the concept of SC, ERC and IRC, is described in 4.2.4. However, for sidelit rooms of simple configuration, where overcast skies prevail, it is appropriate to use the average daylight factor equation[3] to determine the approximate area of opening (W) required to achieve a given average target daylight factor (DF_{ave}). This is useful as a first approximation at the strategic design stage (Figure 4.24).

$$W = [DF_{ave} . A . (1 - R^2)] / [d . T . M]$$

where:

W = opening size (glass area only) (m^2)
DF_{ave} = average daylight factor (%)
A = area of all interior surfaces (m^2)
R = average interior reflectance
d = vertical angle of unobstructed view of sky from centre of window (in degrees)
T = transmittance of glazing
M = maintenance factor of glazing (see Table 11.5)
$R = (a_1r_1 + a_2r_2 ...) / (a_1+a_2...)$
a = area of surface
r = reflectance of surface

Some readers may be bothered by the rather curious use of the vertical angle measured in degrees (and not radians) without a trigonometric function. This is because a constant

$\dfrac{2\pi \times 100}{360}$, is approximated to 1.

This approximate formula illustrates that the average DF

is proportional to the area of the glazing. This is true for areas of glazing above the workplane height.

In the sizing of openings it is not only the issue of daylighting that should be considered, but also the effects of heat loss, solar gain, and the nature of the views out.

4.3.5 The shape of windows

It is common for the tall windows of the 18th and 19th century to be praised for their good daylighting properties (Figure 4.25). In contrast, the modern trend for continuous ribbon glazing with reduced window head height, and often lower ceilings, certainly has caused problems of poor daylight distribution. The key issue is daylight penetration to the back of the room in relation to the illuminance in the front of the room, and as shown in 4.3.4 above this is governed largely by the elevation and proportion of the glazing area that is above the workplane. The shape itself is not of primary significance, except in so far as it influences the zone on the window wall occupied by glazing.

This is illustrated in Figure 4.26, which shows the daylight distribution from two tall windows, compared with horizontal strip glazing of the same area. The tall windows provide better penetration because part of their area is high up on the window wall. Note that the variation across the room (parallel to the window wall) is much less than from front to back. This pattern of glazing also carries the advantage over the horizontal ribbon that it provides an area of shade enabling occupants to avoid the sunpatches.

The graphs (Figures 4.26 and 4.27) show the illuminance levels for the CIE overcast sky. The vertical scale represents illuminance. The horizontal axes represent the position on the 'workplane', a plane at 0.75 m above the floor. The units are metres, measured from one of the window walls. The logarithmic graph hides the absolute difference in illuminance at the back of the room.

4.3.6 Multiple-aspect rooms

If it is possible to provide glazed openings in more than one wall of a room, the problem of uneven illuminance distribution is greatly reduced. Figure 4.27 shows a series of cases where a glazing area of 15% of the floor area is distributed on one, two, three and four walls of the rectangular room respectively. Table 4.4 shows the minimum and maximum illuminances, their ratio and the mean.

Figure 4.25 The tall narrow windows of the Georgian style provide a significant sky component deep into the room.

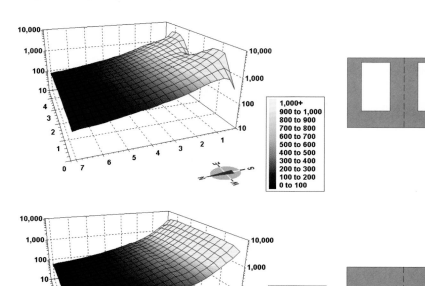
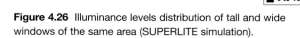

Figure 4.26 Illuminance levels distribution of tall and wide windows of the same area (SUPERLITE simulation).

This shows a slight improvement with the increasing number of apertures, but it is deceptive. More important than the absolute minima and maxima is the prevailing condition, and this is nicely shown on the graphical representation. Note that, for the multiple aspect rooms, a much greater proportion of the room remains in the middle range of illuminance.

It is important to note that multiple-aspect solutions do not have to be symmetric. Even a small contribution of

Figure 4.27
Illuminance levels distribution for multiple-aspect rooms (SUPER-LITE simulation). In all cases the total size of the windows is 15% of the floor area.

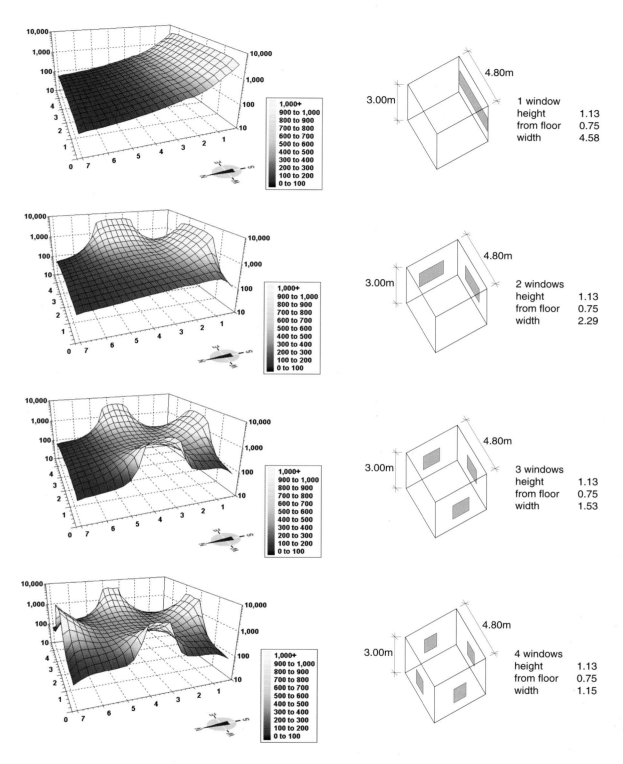

daylight at the back of a room can usefully reduce the contrast. Figure 4.28 shows where in the second case a small window has been added to the back of a single aspect sidelit room. Since the DF is small, a small contribution of extra light can make a significant difference, in this case halving the max/min ratio. This small but valuable contribution of light could be provided by secondary light from another room, atrium or lightwell, as in the case of the Modane School (Figure 4.2).

Table 4.4: Daylight distribution for multiple-aspect rooms.

Windows in facade	Minimum illuminance (lux)	Maximum illuminance (lux)	Ratio	Mean (lux)
S	65	5270	81	705
S & W	59	5213	88	731
S & N	87	5171	59	726
S, E & W	71	4849	68	707
S, E, W & N	69	4596	66	696

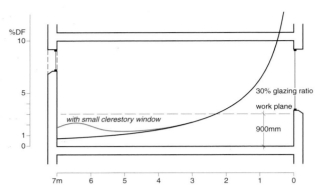

Figure 4.28 The addition of a small window at the back of the room doubles the DF there and thus halves the max/min ratio from 22 to 11.

4.3.7 Effect of room reflectances

The important role of reflected light has already been discussed in section 4.2. It was pointed out that whilst the sky component and the externally reflected component both diminish rapidly with distance from the window, the internally reflected component remains approximately constant.

This is illustrated in Figure 4.29, which compares four rooms of different reflectance. It is easy to see a very significant improvement to the lighting conditions at the back of the room for the most reflective finishes. This is also indicated in Table 4.5.

It follows from the data for the 0, 0, 0 case (all black) that the DF at the back of the room due to the SC is only 0.15%, 3.3 times less than the IRC in the case of the medium reflectance (15, 45, 60) room (there is no externally reflected component).

The distribution of surface reflectance is also important. In cellular buildings, with room widths between 1x and 2x the floor to ceiling height, there are sufficient areas of wall providing vertical surfaces, which can be illuminated directly from the sky. If these walls are of light colour, then they will become secondary light sources providing a valuable IRC to the back of the room. The wall reflectance will then be critical.

For open plan spaces there are few vertical walls, and the main reflecting surface is the ceiling. However, the ceiling does not get a direct view of the sky and can only be illuminated by light from below the horizontal. Most of this will be from outside, reflected from the ground, and is thus dependent upon the ground reflectance. This may to some extent be in the control of the architect as part of the landscape design. These effects are summarised in Figure 4.30.

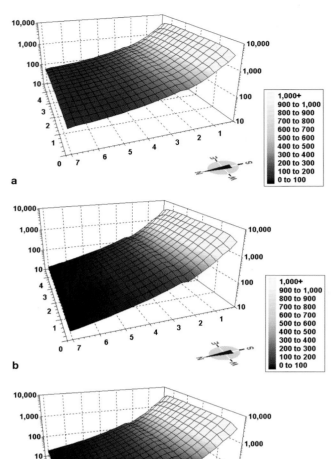

Figure 4.29
Illuminance levels in rooms with different surface reflectances:
(a) reflection coefficients = floor 15%, walls 45%, ceiling 70%;
(b) reflection coefficients = floor 0%, walls 0%, ceiling 0%;
(c) reflection coefficient = floor 5%, walls 10%, ceiling 20%;
(d) reflection coefficients = floor 35%, walls 70%, ceiling 80%.

Table 4.5: Daylight distribution for rooms with different surface reflectances.

Reflectances: floor, wall, ceiling (%)	Minimum illuminance	Maximum illuminance	Ratio	Mean
0, 0, 0	15	5167	344	592
5, 10, 22	22	5184	235	609
15, 45, 70	65	5270	81	705
35, 70, 80	158	5394	34	864

Light may also reach the ceiling from internal horizontal surfaces. These surfaces (floors and tables etc) tend to be dark. If high reflectances are adopted, then in most cases it is even more important that direct sunlight does not fall on them since this will cause serious glare problems. Even outdoor surfaces may cause glare problems, but for normal sill heights these surfaces would not be visible from a sitting position a few metres away from the window.

Figure 4.31 The waffle slab ceiling at the APU Learning Resource Centre, UK, absorbs the low-angle light.

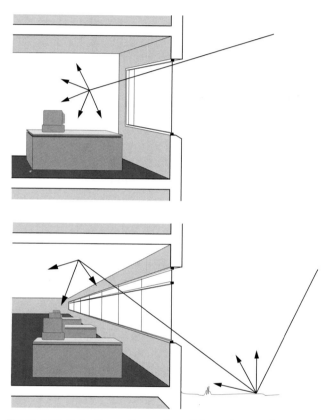

Figure 4.30 Reflection pathways for cellular and open-plan rooms.

Where the ceiling has a role as a reflective surface, it is important to avoid profiles such as downstand beams or coffers running parallel to the window wall. These will become self-shadowing, redirecting light back towards the window, or absorbing it by multiple reflection (Figure 4.31). On the other hand, coffers, which run at right angles to the window wall, do not significantly reduce the reflected light, and may carry other advantages, such as improving the natural ventilation.

4.3.8 Rooflighting

In principle, many of the problems of light distribution resulting from sidelighting can be avoided by admitting light through the roof. Rooflighting (also referred to as toplighting and zenithal lighting, can be located any-

where over the workplane, and thus the depth of plan is no longer a limitation. It is obvious, however, that only single-storey buildings or the top floors of multi-storey buildings can be rooflit.

The simplest form of rooflight is a horizontal aperture in a horizontal roof. However, rooflights take on much more complex geometries than this. There are two functional reasons why the aperture may not be horizontal:

- The plane of the aperture may follow the plane of the roof.

- The plane of the aperture may be deliberately inclined, with a specific orientation, to control the ingress of direct sunlight and/or influence the distribution of light within the space.

A further characteristic is that the surface of the glazing does not necessarily have to follow the plane of the aperture; this will have only a slight effect on the light transmission properties (Figure 4.36).

Figure 4.32 shows a typology of rooflights (using a nomenclature adopted by CIBSE). Note that types 1, 2 and 3 (flat, domes and shed) are following the plane of the roof, whereas types 4, 5, and 6 (northlight, sawtooth and monitor) can be considered as separate structures from the roof element. However, where rooflights of this type occur as repeated elements over a deep-plan building, even these more complex profiles are integrated into the structure of the roof. This is particularly common in large industrial buildings, particularly those built earlier in the 20th century, when the provision of daylight was of high priority.

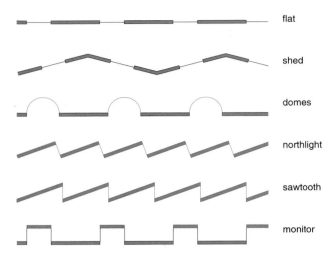

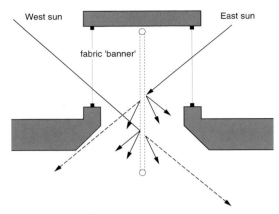

Figure 4.34 East-west facing monitor rooflights can be shaded with vertical reflective/translucent screen.

Figure 4.32 Rooflight configurations according to the CIBSE nomenclature.

All types except flat and dome are orientation-sensitive and this provides an opportunity for solar control. Flat and dome types will almost always need shading from direct solar radiation, which will be at a maximum in the summer on the horizontal. An elegant solution at Stansted Airport is shown in Figure 4.33, where a perforated screen (approx. 60% opaque) reflects some of the light onto the underside of the roof, thus reducing the contrast at the edge of the horizontal aperture. The shading of rooflights is discussed in more detail in section 7.2.

Obviously the name 'northlight' derives from its orientation, although this profile could take on any orientation. Clearly, orientations away from north will admit increasing amounts of direct radiation. If these orientations are adopted, then additional shading devices will be required in most cases.

Figure 4.33 Perforated screens at Stansted Airport, UK (approximately 60% opaque) transmit and reflect light up onto the underside of the roof, providing shading and a reduction of contrast at the edge of the aperture.

The sawtooth roof has similar characteristics to the northlight when orientated with the glazing facing north. However, the vertical plane of the aperture can be easily shaded from summer sun by overhangs. This means that a southerly orientation may be suitable where winter solar gains are required, although the position inside the building, and the nature of the surface where the sunbeams finally fall, must be considered to minimise glare.

The monitor rooflight has vertical glazing in two orientations. As with buildings, orientation north and south is preferable; as in the case above, the south-facing glazing can easily be shaded. If east–west orientation is unavoidable, a single vertical shade can be quite effective in restricting sunlight penetration, as illustrated in Figure 4.34.

Although classified as rooflights, all three types 4, 5 and 6 provide light with a strong horizontal component. In the case of the northlight and the sawtooth this light is unidirectional; in the case of the monitor, bidirectional. This characteristic is often used to advantage in buildings where illumination on a vertical surface is needed, such as picture galleries.

Reflection paths

For the flat, shed and dome rooflights the provision of reflective surfaces has an important role to give a horizontal component to the otherwise predominantly downward direction of illumination. The flat, shed and dome rooflights have no such surfaces other than the reveal in the thickness of the roof structure. This can be extended above or below the roof/ceiling surface. The ratio of the depth of the reveal to the horizontal width of the aperture, and the reflectance of the reveal, are important parameters affecting the light-distribution function.

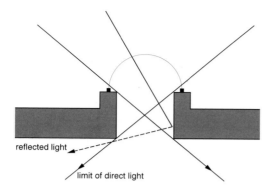

Figure 4.35 The depth to width ratio and the reflectance of the reveal effect the divergence of downward light from the rooflight aperture.

If the surface has high reflectance, > 0.7, the lateral component will generally be improved for reveal to aperture ratios up to about 1.0. If the surface reflectance is low, < 0.3, then the reveal serves only as a restriction to angular divergence of light.

A further function of the reflective reveal is to provide a surface of intermediate luminance between the sky and the ceiling soffit, thereby reducing contrast and hence glare. These functions are illustrated in Figure 4.35.

The northlight, sawtooth and monitor rooflight already have a strong lateral component, but in the two former cases this is unidirectional. If the purpose is to illuminate a predominantly horizontal workplane, then the inclined surfaces in the rooflight can provide reflective surfaces that when illuminated act as secondary sources. This is illustrated in Figure 4.36.

Rooflight spacing to height ratios

The degree to which the downward light from the rooflight diverges will affect the uniformity of illumination on the work plane. The following criteria for illuminance uniformity are given in the *CIBSE Window Design Applications Manual* [1]:

1. minimum/average   0.8
2. minimum/maximum   0.2

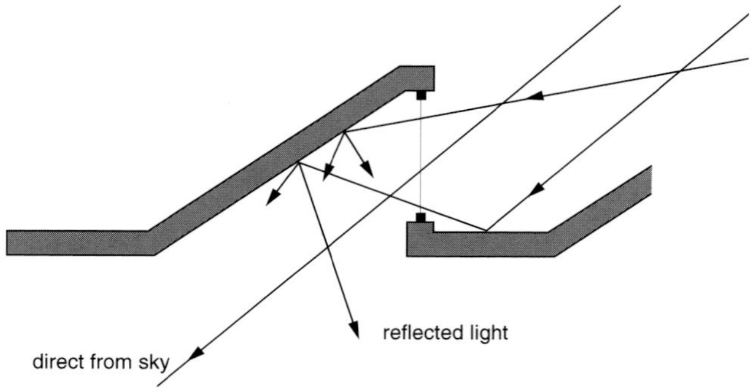

The first is a rather severe criterion, more usually applied to artificial lighting. The second allows for a much larger variation, more likely to result from daylighting.

From model studies carried out in an artificial sky (CIE Overcast), Dewey and Littlefair[4] give the following spacing to height ratios to meet the CIBSE criteria:

Table 4.6: Spacing to height ratios to meet the CIBSE uniformity criteria.

Roof type	Required spacing:height ratio	
	Criterion 1	Criterion 2
Flat	1.5	3.5
Domes	1.5	2.5
Northlight	2.0	3.5
Vertical sawtooth	1.5	3.5
Monitor	3.0	4.5

Note: The northlight and monitor perform better by both criteria, indicating their broader light distribution property.

Approximate daylight factor for rooflights

For a uniform sky, the DF under a flat or dome rooflight system would approximate to the ratio of aperture area to total area of the roof, with a factor to allow for transmittance of the glazing being less than 100%. This would be different for the CIE sky, which has a luminance at the zenith of 3x greater than that at the horizon, and a small correction factor can be applied.

For other types of rooflight further correction factors to take account of the part of the sky visible through the rooflight can be derived and applied to a simple glazing ratio.

This forms the basis of data published in Pilkington's trade literature, represented graphically in Figure 4.37.

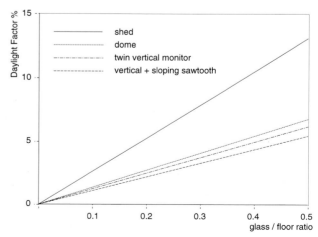

Figure 4.36 Reflected light from the underside of the sawtooth rooflight section reduces the unidirectional transmission characteristic of this configuration.

Figure 4.37 Average daylight factor for rooflights under a diffuse sky. (Pilkington)

4.3.9 Orientation and sunny skies

The orientation of the room affects both the quality and quantity of natural light. South light, as outlined above, is dynamic and generally more intense as well as having different colour-rendering qualities. North light is typically more constant and cooler, and thus particularly appropriate for spaces that require high daylight levels without the risks of overheating or glare.

Even without the contribution from the direct sunlight, the southern sky is brighter than the other orientations, due mainly to the circumsolar zone, that is the area immediately around the disk of the sun. This departure from the cylindrical symmetry of the CIE sky has already been described in 3.2.1.

We can identify three primary considerations for daylight design in relation to direct sunlight:

1. the design of windows to minimise the detrimental effect of glare and unwanted thermal gains

2. the interception and redistribution of direct sunlight to make a useful contribution to daylighting

3. the contribution that externally reflected light from sunlit surfaces makes to useful daylighting in the room

The control of glare and unwanted gains is achieved first by consideration of orientation (3.2) and by shading, which is dealt with in 7.2. The use of direct sun by redistribution is essentially an advanced daylighting technique and is described in detail in Chapter 9. The contribution of externally reflected light from sunlit surfaces is a relatively new area of study, and this too is discussed briefly in 4.2.2.

4.4 Daylight modelling techniques

The prediction of the performance of a proposed day-lighting design is a natural stage in its development. The iterative loop – design proposal – performance prediction – redesign – etc occurs in all design although the depth to which performance prediction and assessment takes place may vary widely. In daylighting design, we have already met some simple prediction tools, such as the Lynes average daylight factor formula (4.3.4), and the Pilkington graphs (4.3.8) for rooflights.

In this section we briefly review three types of approach:

• simplified manual tools
• physical modelling
• computer-based models

Having predicted the daylight characteristics of a design we have to be able to check this against some criteria. Some of these may be quite specific, relating to illuminance variation, such as the Lynes criterion (4.3.3), or glare indices discussed later in 10.1. More detailed criteria and targets for design purposes are given in Part 3.

An answer to the question 'what makes a well daylit building?' would certainly include the absence of defects generically described as glare. It would also include the notion of daylight sufficiency, that is the degree to which there is sufficient daylight to avoid the use of artificial light during the occupied period. This criterion relates directly to the potential energy saving of daylight, and forms the basis of the daylight performance index (DPI), which is also described later in 4.4.4.

4.4.1 Simplified manual tools

Manual tools, that will be of use in daylighting design fall broadly into two categories:

• Tools for predicting daylight factor
• Tools for predicting solar geometry

Daylight factor

Probably the most commonly used tool is the daylight factor protractor. This can be understood by referring to the original definition of the daylight factor, i.e. the ratio of internal horizontal illuminance to external horizontal illuminance. If we were considering an unglazed aperture and a uniform sky, this would be the same as the ratio of the solid angle subtended by the window to the solid angle subtended by the whole sky, i.e. the whole sky dome. The BRE daylight factor protractors in effect measure this solid angle by measuring the angular height and width of the window. However, instead of being calibrated in degrees, they are calibrated directly in daylight factor. The calibration also takes account of the transmission of glass and its variation with angle of incidence (6.1), and the CIE overcast sky luminance distribution.

Figure 4.38 illustrates the use of the two protractors, the first to measure the vertical angle and the second to measure the horizontal angle subtended by the

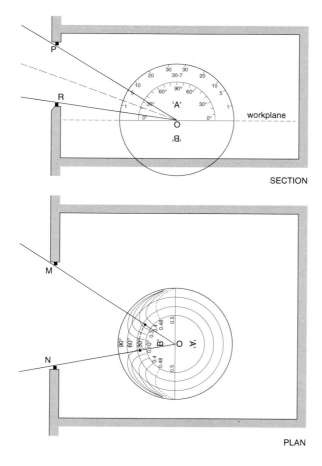

Figure 4.38 BRE Daylight factor protractor in use to measure the sky component for 'wide' windows. The initial SC is then corrected using the other half of the protractor to take account of its finite width. See also 11.6.[6]

SECTION

PLAN

window. The protractors initially measure the sky component but can also be used to measure the external reflected component. The BRE also publishes a nomogram to calculate the internally reflected component. Four pairs of protractors and the IRC nomogram are printed in Part 3 by kind permission of the BRE.

Another tool produced earlier is the Waldram Diagram, shown in Figure 4.39. The diagram represents the projection of half the sky vault. Thus the sky component can be calculated from the ratio of the area visible through the window to twice the total area of the diagram. Note that the vertical angular scale is distorted to take account of the non-uniform CIE sky. The vertical edges of windows or obstructing buildings are plotted vertically, but horizontal edges are plotted as curves interpolating between the guide lines known as 'droop lines'.

Yet another approach to the same problem is the so-called "pepper-pot" diagram suggested by Pleijel[5] (Figure 4.40). Here the sky is represented in stereographic projection and the weighting of the daylight contribution is indicated by the density of dots, each dot representing 0.1% sky component.

Tools for predicting solar geometry

Sunpath diagrams have already been introduced in section 2.1 and Figures 2.2 and 2.3. The cylindrical projection represents a panoramic view with the azimuth angle (compass direction) on the horizontal axis and the elevation from 0 to 90° on the vertical axis. In the circular stereographic projection the circumference represents the horizon and the centre of the circle the zenith of the sky.

Sunpath diagrams have two main uses: to assess the obstructing effect of surrounding buildings and landscape features (Figure 4.41), and to predict the penetration of sunlight into a building (in many cases checking the performance of shading devices).

Sunpath diagrams are presented together with instructions for use in 11.7.

4.4.2 Physical models

The daylighting design of a room can be checked using a reduced scale model. The short wavelength of light (< 1 millionth of a metre) means that the behaviour of light is accurately reproduced at scales down to those likely to be used for architectural models where the smallest dimensions are of the order of 1 millimetre.

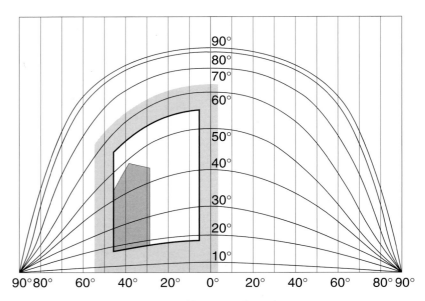

Figure 4.39 The Waldram Diagram. The whole diagram represents a projection of half of the sky vault. The sky component is the ratio of the area of sky visible through the window to twice the total area. The vertical axis is drawn to take account of the non-uniform CIE sky and the variable transmission of the glass with angle of incidence.[7]

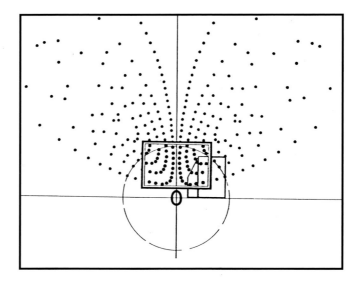

Figure 4.40 Pleijel's 'pepper-pot' diagram. Each dot represents 0.1% SC. The advantage of this diagram is that the geometry is not distorted. The obstruction due to a building can be seen through a window.[8]

Models, usually made of card, plastic or plywood, typically at scales ranging from 1:20 to 1:100, are usually viewed in an artificial sky, a room providing illumination that has the same characteristics as a standard design sky, usually the CIE sky. An advantage of the technique is that both quantitative and qualitative information can be gained. Daylight factors can be measured using photocells and compared with target values and distribution

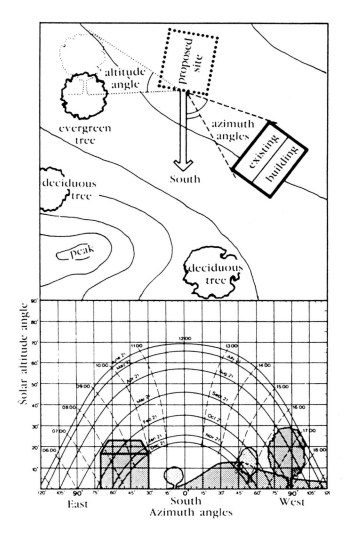

Figure 4.41 The skyline as seen from the point of interest plotted on the cylindrical projection shows the times of day and year when the sun will be obstructed.[9]

criteria. The subjective qualities of the space can also be assessed since the realistic illumination of the model creates a high degree of realism, even though the model may not be highly detailed (Figure 4.42).

Models may also be illuminated under a real sky. Usually this is when sunny conditions prevail as a design condition, since direct sun is difficult to model simultaneously with diffuse daylight in an artificial sky.

Physical models can also be used for sunlighting studies. The model can be illuminated with a spotlight from a particular angle of elevation and azimuth (direction) corresponding to a specific time of day and season, and for the appropriate latitude. These parameters can be determined either from sunpath diagrams or from solar tables. A tilting and rotating table device known as a heliodon allows a range of times and seasons to be investigated quickly and conveniently.

Artificial skies and the technique of physical model testing as a design tool are described in detail in 11.4.

4.4.3 Computer techniques

Computer tools for daylight design fall into two broad categories:

- tools with quantitative output, usually using simple algorithms to predict daylight factor or light transmittance.

- simulation models that use ray-tracing techniques to model the behaviour of light. These techniques can produce highly realistic images as well as quantitative data.

The first type require relatively little input data and are in effect carrying out what the manual methods do, but more quickly and conveniently. For example, the Anglia Polytechnic University DAYLIGHT program requires an input of room dimensions, window dimensions and positions, and surface reflectances. The output includes a DF contour map and synoptic parameters such as average DF and maximum to average ratio, for the CIE sky.

The second type require much more detailed data, including sky luminance distribution, which can in principle be real measured data or average or design skies. The skies can include the sun, making them suitable to model the effect of shading devices and advanced daylighting systems. RADIANCE is probably the most powerful and best known of this type and has been used in the Daylight Europe simulation studies (Part 3). It can produce highly realistic images, as shown in Figure 4.43. Computer-based tools are described in more detail in Chapter 11.

4.4.4 Daylight performance index

In monitoring 60 buildings as part of the Daylight Europe (DLE) case studies, it becomes apparent that some objective method of evaluation is desirable. Indeed it is difficult to present case studies for the purpose of promoting good daylight design without offering some kind of judgement. The more objective this judgement can be the better, not only because designers may quite reasonably feel doubtful of a subjective judgement, but more importantly, because an objective procedure of assessment can be used by expert and non-expert alike.

Figure 4.42 European Law Centre, Sounion, Greece. A physical model for daylight measurement and subjective assessment.

What do we mean by a 'well daylit building'? At its simplest, it might appear that the best daylit building was one that provided a maximum amount of daylight – i.e. the maximum floor area lit for the maximum amount of time. But we know that this strategy would lead to very large areas of glazing, which would result in problems of glare, poor illuminance distribution, excessive solar gains in summer, and excessive heat losses in winter. In response to this, an index was required to penalise over-illumination in some way, balancing the disadvantages to the over-lit zone with the advantages to the marginally illuminated zones.

As part of the Daylight Europe project a daylight performance index (DPI) has been defined. It establishes a daylight factor contour of minimum daylight sufficiency, and a daylight factor contour of over-illumination. Only the area of the building between these two contours counts positively towards the DPI. The value of the minimum daylight factor is determined from the prevailing regional sky luminance climate and an internal datum illuminance appropriate to the building use. The value of the maximum daylight factor is simply

Figure 4.43 Images generated by the ray-tracing model RADIANCE. In (a) illuminance values are shown in false colour. In (b) the same scene is shown with realistic luminance rendering.

four times that of the minimum value. The concept is illustrated in Figure 4.44.

So far, the DPI has been developed only for diffuse skies. The performance of the design in controlling direct sunlight penetration has to be tested separately. Furthermore, in sunny climates where direct radiation may make a useful contribution to daylighting, the DPI described here is not appropriate. The DPI is a relatively new concept, and is presented here as such rather than as a tried and tested tool.

Figure 4.44 The concept of the daylight performance index. The minimum DF contour is the boundary of daylight sufficiency defined as providing the datum illuminance (e.g. 200 lux) for 70% of the occupied period over the year. The maximum DF indicates the area of over-illumination. The DPI is the area between the DF contours as a fraction of the total area of the room.

Summary

Useful daylighting can usually be achieved over a distance equal to twice the floor to top of window height.

The daylight factor (DF) of a room can be considered to be made up of three components – the sky component (SC), the externally reflected component (ERC) and the internally reflected component (IRC).

The SC is proportional to the area (solid angle) of unobstructed sky visible from the reference point. The ERC is proportional to the product of the area of the obstruction and the reflectance of the obstruction. It is important in dense urban sites where the sky may be obstructed by other buildings. The IRC is dependent on the reflectance of the room surfaces and the area of glazing. The IRC is important in reducing the variation of DF.

For sidelit rooms, whilst complete uniformity is not desirable, a ratio of less than 3 for the average DF in the brightest half of the room and the average DF in the darkest part of the room is a good target. Room characteristics that will help achieve this are as follows:

- windows located high in window wall

- windows well distributed (on different walls if possible)

- high-reflectance finishes

- freedom from external obstructions

Rooflighting is very effective since, irrespective of the orientation and inclination of the rooflight aperture, daylight can be delivered anywhere over the floor plan.

The DF distribution of a proposed design can be tested with a range of tools including manual daylight protractors, computer based calculation methods, and physical models.

References

1. *Window Design*, CIBSE Applications Manual (London: Chartered Institution of Building Services Engineers, 1987).

2. J A Lynes and P J Littlefair "Lighting energy savings from daylight estimation at sketch design stage" *Lighting Research and Technology* 22 (3), 1990, pp. 129–137.

3. *Window Design*, CIBSE Applications Manual (London: Chartered Institution of Building Services Engineers, 1987).

4. P J Littlefair and E J Dewey, "Rooflight spacing and uniformity", *Lighting Research and Technology*.

5. G Pleijel "The Computation of Natural Radiation in Architecture and Town Planning" Meddelande Bulletin No. 25. Stockholm: Stateus Nämnd för Byggnadsforskning.

6. R G Hopkinson, P Petherbridge and J Longmore, *Daylighting* (London: Heinemann, 1966), p. 133.

7. *Ibid*, p. 181.

8. Pilkington, *Windows and Environment*, p. 2.9.

9. J R Goulding, J O Lewis and T C Steemers, *Energy in Architecture: The European Passive Solar Handbook* (London: CEC/Batsford, 1992), p. 5.9.

5 Refurbishment of existing buildings

Refurbished lightwell
in Seville at the
home of artist
Santago del Campo.
Architect:
Lopez de Asian

5.1 Daylighting options

The rate of replacement of old buildings with new buildings is very slow, less than 2% per annum for most European countries. Thus methods for improving the quality and use of daylight in existing buildings are of great importance, both from the viewpoint of an improved working environment, and for energy savings. Work on existing buildings covers a spectrum from internal fit-outs of furniture and fittings to major re-modelling involving replacement of facades and even partial demolition.

Refurbishment offers a number of possibilities in improving the lighting design and energy efficiency, depending on where on the spectrum of refurbishment the case lies. These fall into three related categories:

1. the availability and quality of daylight in the room

2. the use of daylight (displacement of artificial light)

3. the efficiency with which artificial light is provided.

The availability and quality of daylight in a given building can be improved by:

- increasing the size of existing apertures or making new apertures, or redistributing the glazing (4.3)

- increasing the transmittance of windows by reducing obstruction due to framing or replacing the glazing material with one of higher transmittance

- increasing the external reflected component by treating nearby external surfaces with high-reflectance finishes

- increasing the internal reflected component by treating internal room surfaces with high-reflectance finishes

- increasing the penetration of light using special elements such as lightshelves or prismatic glazing (Chapter 9).

The use of daylight may be improved by:

- improving the daylight distribution in the room – i.e. reducing contrast to meet the Lynes 3:1 criterion (4.3.3).

- installation of a photosensitive control system (8.1)

The efficacy of the artificial lighting can be improved by:

- re-lamping with high-efficacy light sources

- improving the light output (light output ratio and the utilization factor) of the luminaire

Most of the principles involved in improving daylighting as part of refurbishment are common to new-build and are therefore dealt with in other sections of this book.

5.1.1 Changes to existing apertures

In cases where there is insufficient glazing area and an opportunity to increase the area as part of a remodelling, it is rarely simply a case of increasing the area alone. The opportunity should be taken to increase the area where it is most effective. This will nearly always result in increasing the height of the windows. This will improve daylight penetration to the back of the room, where the critical condition exists. Whilst not related to daylighting, the provision of openable windows at high level also has benefits for ventilation.

This point is made in Figure 5.1, which shows the DF profile for three window options of the same area but different shape: wide, medium and tall. Between 3 m and 5 m from the window, the tall window provides twice as much light as the wide window. Since the illumination in the back half of the room is always the critical factor, this is very significant.

The increase in height of an existing window may sometimes interfere with a suspended ceiling, or may be impossible owing to there being structural members in

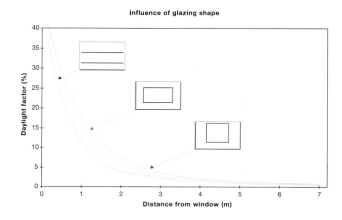

Figure 5.1 Daylight factor profiles for three different window configurations of the same total area. Between 3 m and 5 m from the window, the taller window provides twice as much light as the wide window.

Figure 5.2 Provision of sloping suspended ceiling or partial removal of suspended ceiling to improve daylighting.

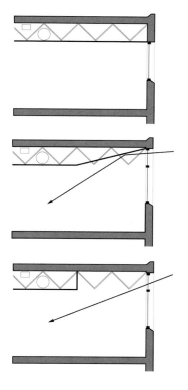

the wall. In the former case, the suspended ceiling can be sloped up to the top of the window or partly removed. The remaining part can provide service space and some acoustic absorption, whilst leaving the structural slab partly exposed as coupled thermal mass for the space (Figure 5.2).

It is also obvious that the installation of a raised floor has the same effect as lowering both the ceiling and the top of the windows, and will be detrimental to daylight penetration.

Alterations to existing apertures can be made to improve the visual comfort. Glazing bars and reveals with dark finishes can be refinished in light- coloured paints. The wall in which the window is located should also be of light colour. These measures will reduce the contrast between the sky and the adjacent surfaces. The splaying of reveals

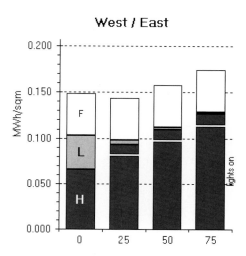

West / East

Figure 5.4 A graph from the LT method showing a reduction of overall energy consumption of 20% by reducing the glazing from 75% to 25% of the facade of a school building. Thermal comfort would be improved, glare reduced and extra wall space provided. The horizontal axis is the glazing area as a percentage of the total wall area.

can also improve visual comfort by providing a surface of intermediate brightness between the wall and the sky (Figure 5.3).

There may often be cases where glazing area can be reduced without leading to a significant increase in artificial lighting energy. Many buildings of the last 40 years are overglazed, and reducing the glazing area can reduce glare and thermal discomfort, as well as reducing heating and cooling loads. This is illustrated in Figure 5.4, which

Figure 5.3 The splayed reveal of this church window provides a surface of intermediate brightness between the sky and the wall in which the window is located, reducing contrast.

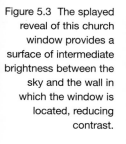

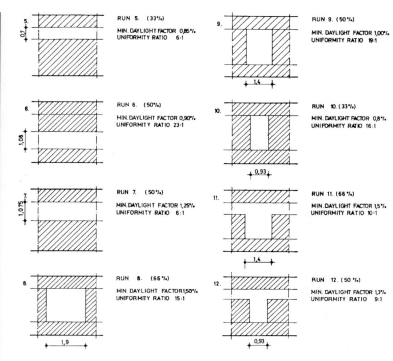

Figure 5.5 Glazing reduction options for typical 1970s system-built school, also showing percentage reduction, minimum DF and uniformity ratio (ECD Partnership).

Figure 5.6 Daylight model of proposed refurbishment to SCOLA System school during refurbishment by architect Edward Cullinan. Shows reduced glazing area and addition of rooflight to improve daylight distribution.

Figure 5.7 Remodelling of the Plaza da Americas on the EXPO 92 site (Architect: Lopez de Asian). Four lighting courts (or patios) were cut into the deep-plan building to provide light and air to the surrounding spaces.

shows a graph from the LT method (8.2.7). Considerations must be made about the effectiveness of the glazing, removing that which is least effective. Figure 5.5 shows a range of options for reducing the overglazed facade of a system-built UK school built in the 1970s. The two options (lower right) show the best compromise between view and uniformity. Figure 5.6 shows a model of a proposed refurbishment to a school in the UK, being tested in an artificial sky.

5.1.2   Making new apertures

Any alteration to the fenestration area will have structural implications. Increasing areas of existing glazing by increasing the height may often be impossible, and a better solution might be to make new, separate apertures. This gives the opportunity for positioning them where they are most effective, and as a general rule this will be where they reduce the variance – i.e. they increase the daylight in the darkest part of the room preferentially.

This will be much more effective if the new apertures are not in the existing window wall, but in an adjacent or opposite (back) wall. It is important to note that the value of the extra window is not dependent upon the area being as great as the existing windows. A modest improvement of the DF at the back of the room – from say 1% to 2% – is very valuable, since it reduces the DF ratio significantly (see Figure 4.28).

This approach, the improvement of the minimum DF, can justify retrofit measures such as glazing into another

daylit internal space – secondary daylighting. It may also justify the retrofitting of advanced daylighting systems, described in Chapter 9.

In single-storey buildings, and on top floors, rooflights are a very efficient daylighting improvement. This is because of their high yield of light from the diffuse sky, and the fact that they can be positioned anywhere in the plan.

Occasionally, it is possible to carry out major remodelling and bring light deep into a multi-storey building

Figure 5.8 Shakespeare Eastleigh School, Hampshire, UK (Plinke Leaman & Browning). The loss of light due to the roof glazing was more than compensated by painting white the original black fascias.

by cutting out lightwells or atria (3.3) as illustrated in Figure 5.7. Or it may be possible to improve the daylight performance of an existing lightwell or courtyard. Figure 5.8 shows an atrium formed by covering over an existing courtyard in a school in Hampshire, UK. In spite of the loss of light by the reduced transmission of the roof glazing, the daylighting of the surrounding spaces was improved by replacing the original black finish with white paint.

5.1.3 Increasing transmission of windows

The existing window openings in the opaque envelope may be adequate in area, but of poor light transmittance.

This can be due to obstruction due to heavy framing, or glass of low light transmittance, either by specification, or due to lack of maintenance.

Framing can obscure more light than might be expected. Figure 5.9 shows the proportion of the total area obstructed by framing when viewed at right angles to plane of the window. However, the actual situation is worse than this. The framing has thickness, and this has a further obstructing effect for light striking the glazing at an oblique angle. Since in most cases the majority of the light from the sky comes from at least 30° above horizontal, and often above 45°, this further obstruction can be very significant. It is difficult to generalise, but the percentage obstructions in Figure 5.9 could easily be increased to 11%, 28% and 55% respectively. Clearly framing with the deeper sections is having more effect. A reasonably accurate estimate could be made by drawing a section of the window with 45° 'shadow lines' from the sky, indicating the increased area of obstruction.

Obviously replacing window glazing with more slender framing can improve transmission. Where this is not possible, some improvement could be made by ensuring that the framing is as light a colour as possible. Unfortunately, in smaller windows, modern timber replacements for double-glazed units often have thicker framing than the traditional single-glazed frames.

The glazing itself may have poor transmission. This may be because tinted or reflective glass (7.2) has been specified inappropriately, often as a matter of style rather than function. Sometimes, tinted or reflective films have been added for the same reason, or without regard for orientation. Replacement with clear glass may be feasible as part of improvement of thermal performance by double glazing.

There may be some situations where tinted or reflective glass has been installed appropriately in order to reduce solar gains and glare from direct sunlight. Careful con-

Figure 5.9 The proportion of the total window opening obstructed by the frame.[1]

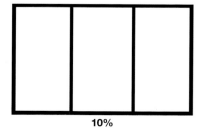

10%

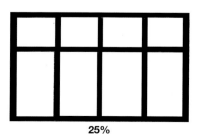

25%

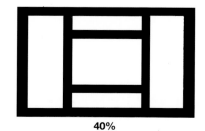

40%

sideration has to be given in these cases, but it might still be possible to replace it with clear glazing together with movable shading devices or shading with a light redistribution function, as described in section 7.2.

5.1.4 Improving light reflection

Reflected light almost always makes an important contribution at critical locations in the room where the daylighting is minimal. This has already been explained in 4.2 and 4.3. Both the internally reflected component and the externally reflected component can be improved as part of refurbishment.

The importance of reflection pathways has been illustrated in Figure 4.30. The repainting of sidewalls with reflective paint and the removal of obstructing furnishings will improve the daylighting of a cellular office, both the illuminance distribution (increasing the uniformity) and improving visual comfort by reducing the contrast at the edge of the window. Some caution has to be taken when increasing reflectance of side walls that may be sunlit. In these cases, measures should include shading devices to prevent sunpatches on the side walls.

In the case of open-plan or wide-plan rooms the critical reflection pathway is the ground and ceiling. Clearly the ceiling can have its reflectance increased as part of redecoration. Improvements may also be made by removing obstructing light fittings and other objects that intercept the light at grazing angle to the ceiling surface (Figure 5.10).

The treatment of external surfaces may also be part of a refurbishment operation. Changing a black tarmac carpark surface to concrete paving of medium reflectance can make a significant difference. In cases where windows overlook flat roofs, upgrading of the roof may include the addition of white stone chips or a white reflective paint for thermal reasons, and this will also improve the daylighting. In cold climates, the effect of snow in improving the daylight distribution is a common experience.

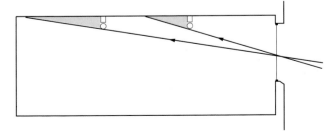

Figure 5.10 Even quite small obstructions such as light fittings can intercept the shallow-angle light, reducing the illuminance of the part of the ceiling away from the window.

Figure 5.11 Access deck at the Bertholt Brecht School, Dresden, Germany. New structures erected as part of refurbishment and remodelling may actually reduce the interior daylight. When obstructions are unavoidable, light-coloured finishes will reduce their impact.

Highly reflective ground surfaces do create an increased risk of glare when viewed directly from a position close to the window. However, from normal eye height when seated, only the distant ground would be visible.

Above-ground obstructions such as buildings or walls may also be treated to improve the ERC. In urban situations, a large fraction of useful daylight will be from this source. This has been discussed in some detail in 4.2. Building plans with internal corners, such as L and T shapes, will create self-obstruction, which can be mitigated by external wall finishes of high reflectance.

Occasionally construction that is part of the refurbishment may worsen daylighting conditions. This could be the addition of a staircase or access deck as in Figure 5.11. As a general rule, where obstruction is unavoidable, treating the surface with a high reflectance finish will always mitigate the effect to some extent.

5.1.5 Shading devices and advanced daylighting systems

In sunny conditions, for rooms that would otherwise receive direct sun, it is not uncommon to observe the blinds fully closed or the curtains drawn, and the lights on.

However, although it may seem to be counter-intuitive, the installation of controllable blinds can actually lead to an increase in the use of daylight. The degree to which shading systems can modulate the incident direct sunlight, so that the sunlight can still provide a useful source of light without glare and excessive solar gain, is discussed in section 7.2.

Advanced systems may provide more than this, increasing the utilisation of direct sunlight by redirecting it to the back of the room. Advanced systems could be an appropriate refurbishment option in sunny climates for this reason (Chapter 9.).

5.1.6 Improvement to artificial lighting system efficiency and controls

Although this is not an improvement to daylighting, improving artificial lighting and controls will also result in energy savings. Indeed, quite modest improvements in lighting efficiency may have much more impact on energy use than more costly improvements to the daylight environment, and should thus be high on the list of priorities. For example, it is quite likely that replacing 10-year-old luminaires with high-efficiency state-of-the-art lamps and luminaires can halve the installed power for the same working illuminance, thereby halving the lighting costs, even if no other measures are taken.

The energy used by lighting systems, and the potential saving by the use of daylight, are discussed in Chapter 8.. The LT method, which allows the impact of refurbishment measures to be assessed, is also described in 8.2.7.

5.2 Summary

Refurbishment provides the opportunity for increasing the lighting performance of buildings broadly in three ways:

1. the availability and quality of daylight in the room

2. the use of daylight (displacement of artificial light)

3. the efficiency with which artificial light is provided

The availability and quality of daylight in an existing building can be improved by:

- increasing the size of existing apertures or making new apertures, or redistributing the glazing

- increasing the transmittance of windows by reducing obstruction due to framing or replacing the glazing material with one of higher transmittance

- increasing the ERC by treating nearby external surfaces with high-reflectance finishes

- increasing the IRC by treating internal room surfaces with high-reflectance finishes

- increasing the penetration of light using special elements such as light shelves or prismatic glazing

The use of daylight may be improved by:

- improving the daylight distribution in the room – i.e. reducing contrast.

- installation of photosensitive control system

The efficacy of the artificial lighting can be improved by:

- re-lamping with high-efficacy light sources

- improving the light output (light output ratio and the utilization factor) of the luminaire

References

1. R G Hopkinson, P Petherbridge and J Longmore, *Daylighting* (London: Heinemann, 1966), p. 101.

6 Materials, surfaces, light and colour

The Breakfast Room of Sir John Soane's Museum, London. The rooflight, glazed with coloured glass, floods the room with warm yellow light. (Courtesy of the Trustees of Sir John Soane's Museum)

Materials, surfaces, light and colour

Our perception of architectural space is mainly by the reflection of light from surfaces rather than the sensing of direct light from luminous sources. Furthermore, much of the light in a room that provides useful illumination for the task is not directly from the source – the sky or an artificial source – but reflected from surfaces. It follows that the properties of surfaces to affect the light that falls on them – the texture, reflectance and the colour are of concern to the architect. Although the mechanical properties of materials – their strength, resistance to abrasion, to fire, effect of dirt and pollution – are all important considerations in design, it is the optical properties that are held in the mind's eye at the time when the design is being conceived. The inherent colour and texture of stone, timber, and brick, and the applied colour of paint, ceramics and textiles, are an essential part of the architect's conceptual palette.

This chapter describes the optical properties of materials – in particular texture and colour. First we shall deal with reflection and the transmission of light without considering the issue of colour.

6.1 Reflection, transmission and absorption

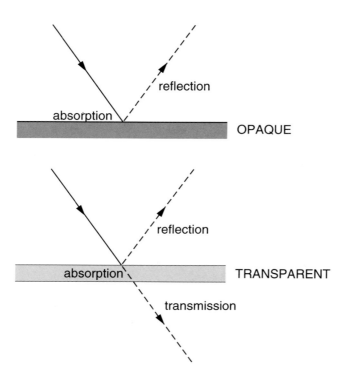

Figure 6.1 The processes occurring when light strikes a surface or layer of material – absorption, reflection and transmission.

There are three possible processes that can take place when light falls onto a surface or a layer (Figure 6.1):

- Light can be reflected – the ratio of the reflected energy to the incident energy is called the reflectance, R.

- Light can be transmitted – the ratio of transmitted energy to incident energy is called the transmittance, T.

- In both cases energy may be absorbed – the ratio of absorbed energy to the incident energy is called the absorptance, a.

Considering reflectance, perfect black would be 0, perfect white would be 1, and all real surfaces lie somewhere in between. For opaque materials the transmittance is zero and thus the absorptance and the reflectance are related as follows:

$$R = (1 - a)$$

or, in words, the reflected energy is equal to the incident energy minus the absorbed energy – i.e. obeying the law of conservation of energy.

The reflectance of some common materials is given in 12.3.

For a transparent material in the form of a layer, the incident energy is reflected, transmitted and absorbed:

$$R = (1 - a - T)$$

and again the total energy is conserved.

Note that for opaque materials the absorption takes place at the surface, whereas for transparent materials the absorption takes place in the body of the material and is dependent upon the thickness of the material.

Typical values for clear 6 mm glass are:

reflectance	0.1
absorptance	0.05
transmittance	0.85

These parameters are often quoted as percentages. Note that reflectance, absorbtance and transmittance can be qualified by total, which means all visible and invisible radiation, or visible.

6.1.1 Specular and diffusing materials

The properties of opaque or transparent materials vary as a function of the direction of the incident light. It is easier to define these properties in relation to a unidirectional (or direct) incident beam, such as the one produced by the sun or a distant point source.

If an opaque surface is said to be specular, it means that it reflects like a mirror – a direct beam is reflected as a direct beam, following the law of reflection, i.e.

angle of incidence i = angle of reflection r

The reflectance is the ratio of the reflected light flux (or energy) to the incident light flux (or energy). We refer to this reflectance as the direct/direct reflectance.

If a transparent material is described as specular it means that it transmits a direct beam of light without dispersing or diffusing it, and it follows from this that a sharp focused view can be seen through the material.

The transmittance is the ratio of the transmitted flux to the incident flux, which again for a specular material would be the direct/direct transmittance.

Both opaque and transparent surfaces (or layers) can be diffusing. The diffusion of light means that the reflected or transmitted light flux is distributed in all directions (towards the surface of an imaginary hemisphere), although the incident beam may be direct (unidirectional).

These four conditions are illustrated in Figure 6.2. Note that the rays are drawn with a length that is in scale with their intensity. The circular envelope to the reflected or transmitted rays indicates that the brightness of the surface will be the same viewed from any angle. This is known as a 'Lambertian' reflector after the scientist Lambert, who was the first to define such a surface. In both the diffusing cases, because the energy is redis-

tributed, the reflectance or transmittance is referred to as direct/hemispherical.

In daylighting, our prime concern is for the diffuse light arriving from the sky vault. Since the light source is 'hemispherical', we shall refer to hemispherical/hemispherical reflectances or transmittances (Figure 6.3).

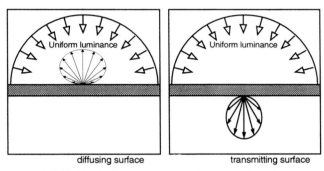

Figure 6.3 Light reflectance and transmittance patterns for a hemispherical incident light source. Typically, hemispherical transmittances are lower in value than directional ones measured at right angles to the surface, by at least 20%, owing to the increased reflectance at low angles of incidence.

6.1.2 Real surfaces

Glossy surfaces, or surfaces covered with a glass sheet or a varnish, add a specular component to a diffusing one, resulting in a hybrid characteristic. The darker the surface, the lower the diffuse reflectance. However, the specular reflection takes place in the outermost layer of the surface, and may not be affected by the pigments and hence may remain constant for light or dark surfaces (Figure 6.4).

Perfect diffusion of light on reflection is rather rare in building materials. Some fabrics, such as carpets, can show almost 'Lambertian' characteristics, but this can be influenced by the lay of the fibres. Even flat paints present non-symmetrical patterns in their reflective proper-

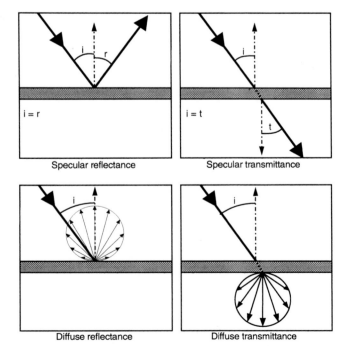

Figure 6.2 Specular and diffuse light reflectance and transmission patterns for a directional incident beam. Note that the length of the small arrows represents the luminous intensity.

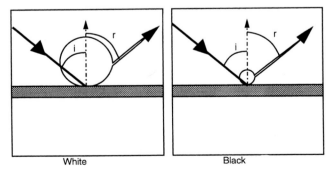

Figure 6.4 Hybrid reflectance patterns for white and black glossy paints. Note that in this diagram the arrows representing diffuse reflection have been omitted for clarity; only the envelope has been shown.

ties. However, in a situation of multiple light reflections in a room, their overall behaviour becomes near to one of a perfect diffuser, leading to this simplification: most indoor surfaces can be considered to be perfectly diffusing for daylight factor calculations.

Partial light- scattering properties may however be of high importance for some predominantly specular materials such as glazing materials or reflectors. They can smooth and soften the specular images, but still keeping roughly the same directionality. Sand-blasted and etched glass or brushed aluminium foils are examples of this (Figure 6.5).

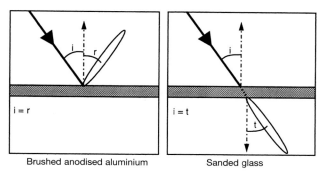

Brushed anodised aluminium   Sanded glass

Figure 6.5 Near-specular behaviour of reflectors and transparent elements (e.g. etched glass).

The reflectance of surfaces of a room has a considerable impact on the penetration of light deep into the room, owing to the effect on the internal reflected component (4.2.3). We can use the simple formula below to calculate the impact of reflectance on the IRC (in lumens):

$$IRC = R \cdot F / S (1 - R)$$

where

S   is the total area of the surfaces of the room
R   is the average reflectance of the surfaces
F   is the luminous flux (lumens) entering the room

Let us assume for a reference room with mean reflectance $R = 0.3$ that the IRC at a point in the back of the room is 200 lux. If we change the mean reflectance we can see a dramatic change in the IRC, as shown in Table 6.1.

Table 6.1: Change of IRC depending on the reflectance of the surfaces.

Average reflectance	IRC (lux)
0.1	52
0.3	200
0.5	466
0.7	1087

Bearing in mind that the IRC often constitutes more than half of the illumination in the most poorly illuminated part of the room, this analysis indicates the importance of reflectance in daylighting design.

The properties of glazing materials play a significant role in the process of daylight penetration in a room. For instance, tinted glazing or diffusing glass often reduces the amount of daylight transmitted by more than 60%, in comparison with clear glass. Indoor reflectances have a direct influence on the illuminances in the darkest areas of a room, as shown above. Furthermore, the darker the wall finishes, the higher the contrast between the luminances near the window and the luminances away from it. The light in a room with dark finishes will also be much more directional, casting strong shadows, which in general will diminish the lighting effectiveness.

6.2 Light and colour

Our experience of colour has many curiosities. Why do the colours that are so vivid under the midday sun disappear by moonlight, although we can still clearly make out form and some detail? Why do some colours look different when viewed by daylight and artificial light, but other colours look the same? And although a white surface can look white in sunlight, northlight and fluorescent light, why does colour film record it so differently?

The mechanism of colour perception is complex, and to be able to specify the architectural use of colour with a sound knowledge of the implications it is necessary to understand a little of the science behind colour and colour vision.

Without getting too deeply into an existentialist discussion, it really is true that, in the absence of light, there is no colour. Colour is indeed in the light, as demonstrated by Newton in his classic experiment. White light consists of a spectrum of colours – red, yellow, green, blue and violet – and the sensation of whiteness is present when there is a combination of these, although the proportions of different colours do not have to be precise values.

6.2.1 Thermal light

All surfaces at any temperature above absolute zero (–273°C) emit electromagnetic radiation, but only when temperatures are above about 1000°C does this radiation become visible. Radiation from a hot surface is emitted owing to the violent thermal motions of the atoms making up the surface. The temperature of the surface of the sun is about 6000°C, and this gives rise to a spectrum that contains a little over half of its energy in the visible region – extending to infrared and microwaves at longer wavelengths and ultraviolet at short wavelengths (Figure 6.6). This is not a happy coincidence – human eyes have evolved to make the best of available daylight.

A candle, an oil lamp and the common light bulb all produce light by this process, although they operate at much lower temperatures. This results in radiation with a much greater proportion of energy in the infrared region, and the visible part of the spectrum has much more red light than blue light. This relationship between the colour of nominally white light and temperature gives rise to the concept

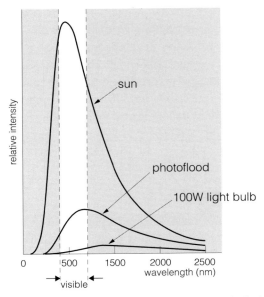

Figure 6.6 The continuous spectrum of the sun. About half of the energy lies in the part of the spectrum visible to the human eye. The spectrum from a high temperature and that from a normal light bulb are also shown.

of 'colour temperature'. Paradoxically, colours that are often described as 'warm', i.e. richer in red light, are in fact a lower colour temperature than so-called 'cool' colours. Colour temperature is dealt with in more detail in 6.4.2.

6.2.2 Electronic light

Hot surfaces are not the only sources of light – energy transitions associated with the electrons inside the atoms can also emit light. The neon tube in which an electric current passes through neon gas at low pressure is a familiar example. The difference between this electronic light and the thermal light described above is that, owing to the precise energy values of the transitions within the atom, light of precise wavelengths or colours is produced. An example is shown in Figure 6.7. This is different from the spectrum of thermal light, such as the solar spectrum, which shows a smoother, more continuous range of wavelengths.

The fluorescent lamp is another form of electronic light. In this case the electric current passes through a gas of mercury vapour and produces ultraviolet light, invisible to the eye. The ultraviolet light then falls on a powder lining the tube, which fluoresces – that is, it absorbs ultraviolet radiation to make electronic energy transi-

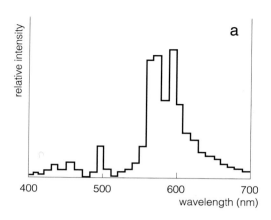

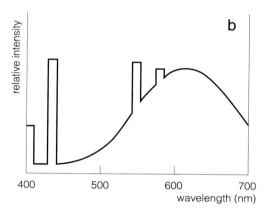

Figure 6.7 Electronic light, from (a) gas discharge sources and (b) fluorescent lamps, has a highly discontinuous spectrum.

- red + green = yellow

- green + blue = blue-green or cyan

- blue + red = magenta

The resultant colours are described as being complementary to the third colour – i.e. the primary colour not present in the combination. Thus:

- yellow is complementary to blue

- cyan is complementary to red

- magenta is complemetary to green

Complementarity simply means that the two colours together provide the sensation of white light. This is nicely demonstrated in Figure 6.8, which shows three overlapping circular beams of light producing primary, secondary and tertiary (or white) light. This can easily be demonstrated with three projectors with colour filters, and is the principle upon which Technicolor and, more recently, video projectors work.

Note that we are discussing here the addition of coloured light, not pigment or dye. Note that the three primary colours are not quite the same as the red, yellow and blue that we were probably told were primary colours in our early art classes. Nor are they the colour of inks for colour printing, which are magenta, yellow and cyan – that is what we are calling secondary colours.

Figure 6.8 The superimposition of two of the three primary colour lights red, green and blue produces the secondary colours yellow, blue-green or cyan, and magenta. The addition of the three primary colours produces white.

tions, and then re-emits it at a longer wavelength in the visible region of the spectrum. By using a mixture of different fluorescent compounds, it is possible to construct a quasi-continuous spectrum. The term 'polyphosphor', often applied to modern fluorescent lamps, refers to this.

6.2.3 Colour sensation

Light from the sun and sky, light from hot surfaces such as candle flames or incandescent filaments, and light from fluorescent lamps all give light that we would broadly describe as white, although, as the spectra show, the composition of this white light varies. We shall now show that the property of colour, which we normally attribute to a surface or object, is due to the absorption of part of the spectrum by the pigment or dye, and the reflection of the remaining parts.

A century after Newton had shown that white light contained all the spectral colours, the scientist Young proposed the tristimulus theory of colour perception. This states that any colour sensation can be generated by a combination of three primary light colours. These were identified as red, green and blue. A combination of any two of these results in a secondary colour sensation:

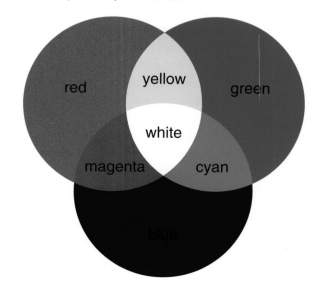

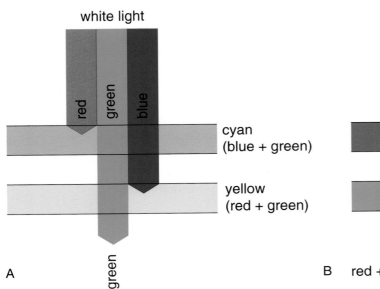

A

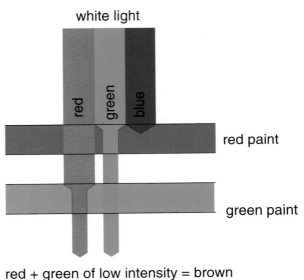

B   red + green of low intensity = brown

Figure 6.9
The addition of colour dyes or pigments is different from the addition of coloured light. In this case we have to consider the addition of absorption: the colour of the light reflected will be that part that neither dye nor pigment absorbs (or which is absorbed the least).

This paradox can easily be explained by reference to Figure 6.9, which shows white light falling onto painted surfaces. For the sake of clarity we have shown the process of transmission rather than reflection, but the principle is the same. In case A white light falls onto a surface painted with a mixture of cyan and yellow. The cyan absorbs the red but transmits the green and the blue (cyan-coloured light contains only blue and green light, as shown in Figure 6.8). We know that the yellow layer will transmit red and green light (as in Figure 6.8) and absorb the blue light. This leaves only the green light to be transmitted, or reflected in the case of an opaque surface.

This is now similar to our experience in mixing paints, although we probably used blue paint instead of cyan. This works because pigments of a nominal colour do not reflect only in that band – it is not a sharp cut-off.

Case B shows the effect of the addition of red and green. These are both primary colours, and in principle the addition of them as pigments should produce black: the red absorbs blue and green, and the green absorbs blue and red. However, as mentioned above, the pigments are not perfectly monochromatic, and both pigments transmit a little of the two other colours. This results in a slight transmission (or reflection) of red and green because only one pigment nominally absorbs each of these, whereas both pigments absorb blue. Red and green light make yellow – but it is of very low luminosity since most of the light has been absorbed by the combined effect of red, green and blue absorbers. We call the colour of dark yellow – brown. Note that although we can perceive red or yellow or white light, we would not refer to brown, grey or black light! These colour names

are reserved for describing the phenomenon of absorption only, i.e. pigments or dyes.

6.2.4   Colour contrast and adaptation

The phenomenon of adaptation (in the context of the perception of light) has already been touched on in 4.1 and will be described in more detail in 10.1.2. It is a familiar experience that, when moving from a brightly lit environment, such as a beach on a sunny day, into a room with relatively small windows (i.e. a small daylight factor), the room at first seems quite dark. Twenty minutes later, the room appears to be adequately lit. This process is fundamental to the operation of the eye, enabling us to see comfortably from about 0.1 lux to 100 000 lux, a range of 1 million fold. This huge range of sensitivity is achieved partly by the change in aperture size of the iris, and partly by a change of the physiological sensitivity of the light-sensitive cells in the retina.

It is less well understood that the virtual red, green and blue sensors show this physiological adaptation independently. This has implications for our perception of white light – when red, green and blue light are present, but not in equal proportion, the eye can create the impression of whiteness by desensitising for the primary colour that predominates. This sophisticated response allows the colour sensation associated with an object to be stabilised irrespective of the colour of the illumination (within limits). Its evolutionary significance is that it enabled our ancestors to use subtle differences in colour to accurately identify non-poisonous fruits and fungi, irrespective of the colour temperature of the daylight.

Modern video camcorders have this same 'white balance' facility; it is possible because as in the eye, it is under 'software' control. The conventional photographic camera does not have this opportunity; colour film has a fixed colour balance sensitivity. When moving from, say, daylight to incandescent light, this colour balance has to be changed either by changing the film or by using a colour filter to change the composition of the incoming light.

The independent adaptation of the primary colour sensors can have more dramatic effects. For example, if subjected to strong light of a magenta colour, the red and blue sensors of the eye become relatively desensi-tised. When the magenta light is replaced by white light, the red and blue sensors will take time to regain their sensitivity; during that time the 'over-sensitive' green sensor will cause a sensation of greenness. This interesting phenomenon is the basis of a number of amusing optical illusions (Figure 6.10), but also accounts for many of the colour phenomena explored by artists and architects. More importantly, it gives a significance to the term 'complementary colour' i.e. the so-called after-image will be in the complementary colour to that of the initial image. This is directly analogous to the photographic colour negative.

Figure 6.10 To demonstrate colour adaptation in the eye, stare with a fixed gaze at one of the stars while under strong light. Then look away towards a neutral grey or white surface. The colours seen in the after-image are complementary to the colours in the original.

6.3 The specification of colour

The description of colour attracts many words to describe their quality, ranging from the everyday – bright, vivid, dull, dark, subtle – to the more extreme – electric, vibrant, lurid, and sickly. Specific colours attract the use of hundreds of similes such as blood red, Eau de Nile, Royal blue, or canary yellow, whilst some names refer quite accurately to the origin of the pigment, e.g. cobalt blue, burnt umber, and crimson lake. Only the latter names have any chance of being truly objective in that they may define an actual chemical substance with a specific colour.

Probably the first person to make a significant step forward in this problem was Albert H Munsell, an American painter born in 1858. At the beginning of his book *A Colour Notation*,[1] first published in 1918, in despair of the then chaotic situation he quotes a line by the author R L Stevenson, "…this red – it's not Turkish and it's not Roman and it's not Indian, but it seems to partake of the two last, and yet it can't be either of them because it ought to go with vermilion …"!

6.3.1 The Munsell colour notation

Munsell was the first to identify three distinct qualities of colour – initially applied to the colour of surfaces but also applicable to light – namely hue, value and chroma, to use his original terminology. The equivalents to these words now in common use are hue, lightness and saturation, but the concept remains the same. The definitions are listed in Table 6.2.

Table 6.2: Munsell values.

Hue	This is the nominal colour – the name we give to the particular sensation e.g. red, orange, yellow, green etc.
Lightness or value	A description of the brightness of a colour on a scale 0 to 10, from dark to light. This may vary within one colour, e.g. dark red – light red, or between colours e.g. yellow (light) – violet (dark)
Saturation or chroma	A measure of the degree of vividness of a colour – from a pure hue, through varying tints of decreasing colour to grey

The way Munsell arranged these qualities, the same in principle as modern systems, is illustrated in Figure

6.11. First the hues were arranged in a colour wheel, recognising that there are intermediate values between the main nominal hues, allowing a continuous cyclic variation of hue. The next step was to represent the chroma (or saturation) on the radius of the circle and the value or lightness on a vertical axis. This leads to the concept of a colour solid as shown in Figure 6.12.

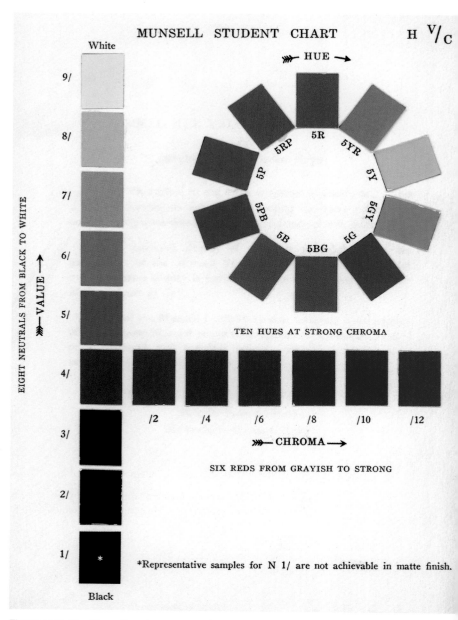

Figure 6.11 The Munsell system arranges hues around the perimeter of a circle and chroma (or saturation) along a radius. Value (or lightness) is on a vertical axis (Munsell Color Company Inc., Baltimore, USA).

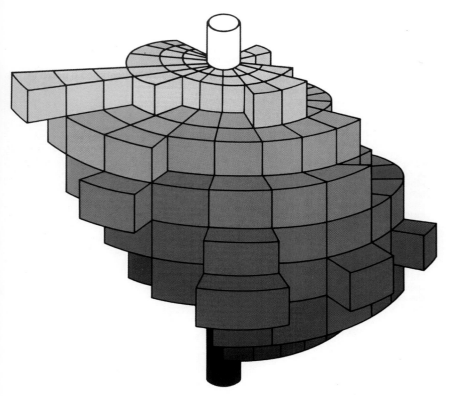

Figure 6.12 Arranging the three variables on a single three-dimensional solid graph leads to the Munsell colour solid (Courtesy of Minolta UK).

In the Munsell system 10 nominal hues are identified:

Red	R
Yellow Red	YR
Yellow	Y
Green Yellow	GY
Green	G
Blue Green	BG
Blue	B
Purple Blue	PB
Purple	P
Red Purple	RP

Each hue has four subdivisions quantified in tenths, e.g. 2.5R, 5R, 7.5R, 10R. The pure or principal hue is denoted by 5, and 10 denotes an intermediate value between the next named hue in a clockwise direction.

Value or lightness is on a 10-point scale from 0 for perfect black to 10 for perfect white (neither is achievable in reality). The value or lightness has a direct relationship to the optical reflectance for white light given by

$$R = V(V-1)\%,$$

where

R is the reflectance
V is the Munsell value or lightness

Thus a colour of Munsell value 8 has a reflectance of 56%. This is the value that could be used in calculating the internal reflected component of a room irrespective of the hue or chroma (saturation).

Chroma or saturation is quantified on a scale from 0, that is no colour or shades of grey, to numbers up to about 18 for the most intense modern pigments. It is interesting to note that in Munsell's original system he regarded 10 as the maximum chroma or saturation.

Using this system any surface colour can be defined by specifying the three colour coordinates. The format is hue/value/chroma, or the hue/lightness/saturation in modern notation. Here are a few examples:

Apple green	7.5 GY	7 / 12
Navy blue	7.5 PB	3 / 4
Ivory	2.5 Y	8 / 2
Neutral mid grey	0	5 / 0

To explain these:

- Apple green has a hue that is on the green side of green/yellow, is quite light and has intense saturation.

- Navy blue has a hue that is on the purple side of purple/blue, has very low value, i.e. is dark, and is of low to medium saturation.

- Ivory is on the yellow red side of yellow, is very light, and of very low saturation.

- Neutral mid grey has no value and is of medium lightness with zero saturation.

The Munsell system is still used to define paint colours in the UK and much of Europe.

Note that the Munsell system defines the colours only by matching – i.e. it is a comparative method based on actual colour samples. A more modern approach is to define the colour of a surface, or light, in relation to the wavelength and the sensitivity of the colour receptors of the eye.

It has already been explained that any colour sensation can be created by a combination of three primary colours. This, the tristimulus principle, forms the basis of modern colour specification systems.

6.3.2 The XYZ tristimulus theory

The concept of the *XYZ* tristimulus values is based on the three-component theory of colour vision. The values of *X, Y* and *Z* are calculated as the product of the incident illumination, the reflectance of the surface, and the sensitivity of the eye for a particular wavelength, and are then integrated over the whole visible spectrum.

For representing this graphically, these values are combined so that chromacity and saturation can be described by just two coordinates *x* and *y*, as shown in Figure 6.13, where:

$$x = X / (X + Y + Z) \quad \text{and} \quad y = Y / (X + Y + Z)$$

i.e. *x* is the strength of the red stimulus relative to the total.

Thus the colour of a red apple, for example, might be described by the point A with colour coordinates *x* = 0.49, *y* = 0.30. The lightness (*Y* value) cannot be specified on the chart but would be given a separate value of, say, 13.0, meaning a white light reflectance of 13%. Because a colour needs three coordinates to describe it – *x,y* and *Y* – the system is referred to as a colour space, and in this case the Yxy colour space.

A more modern development of this system is the *Lab* colour space, where *L* is the lightness coordinate, and *a* and *b* are the combined tristimulus parameters. *a* and *b* are defined differently from *x* and *y* in order to make equal changes in the value of *a* or *b* correspond to equal changes in perceived colour sensation, but the principle remains much the same. The *Lab* colour space is illustrated as a solid in Figure 6.14; chromacity values *a* and *b* take on values from 0 to 60. The *L* value still represents reflectance in %, having values from 0 to 100.

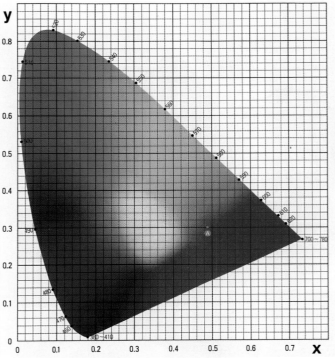

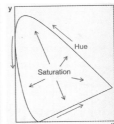

Figure 6.13 The Yxy colour specification system (Courtesy of Minolta UK).

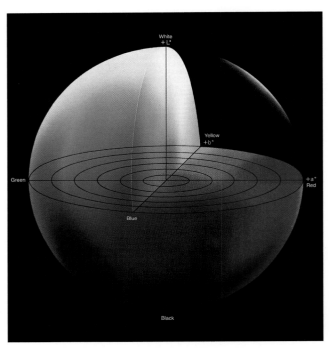

Figure 6.14 Representation of colour solid for the Lab colour space (Courtesy of Minolta UK).

Colour specification is complex, and has taken on a new level of sophistication with the developments in computer graphics. The descriptions above are a brief introduction conveying only the basic concepts of colour specification.

6.4 Colour in architecture

What is the significance of the science of colour to the architect? We can summarise the answer by listing a number of issues:

- reflected light

- colour temperature

- colour rendering

- subjective effects

6.4.1 Reflected light

We have stressed the significance of reflected light, both externally and internally. For example, at the back of a room, the internally reflected component is usually the largest component of the daylight factor and may be the only component in parts of the room. Even reflection from natural finishes (Figure 6.15) can change the colour of the light, although the compensating effect of colour adaptation will often result in this not being perceived. If the main reflecting surface for this light is of strong chroma or saturation, then the reflected light will

Figure 6.15 The relative spectral intensity of daylight, light reflected from concrete, and light reflected from green foliage.

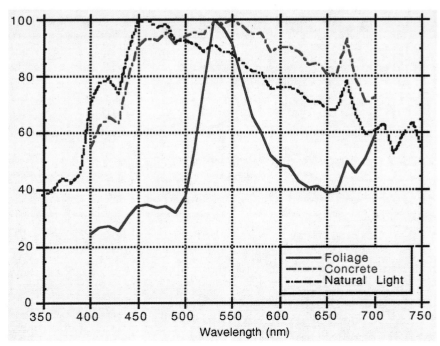

Figure 6.16 School at La Roche sur Yon, France. The saturation of the reflected yellow light is reduced by the white light from the sky and specular reflection from the glossy paint.

take on this character. This may or may not be desirable, particularly if it is a primary colour.

In Figure 6.16, although the rooflight is illuminating a surface of quite strongly saturated yellow, the reflected yellow light is being mixed with white light from the sky. Furthermore, yellow is a secondary colour and thus already contains red and green light. Thus the resultant light in the room will be only slightly biased away from the blue end of the spectrum, which was the intention of the daylight consultant. The maintenance of good colour rendering was of great importance in this instance, a gastronomic teaching kitchen.

Figure 6.17 School at La Roche sur Yon, France. The use of such strong colour is unusual. However, the requirements for colour rendering in the corridor are undemanding.

Figure 6.18 School at La Roche sur Yon, France. The perception of the chroma of the external blue surfaces is enhanced by the complementary yellow finishes of the room interior.

In Figure 6.17 we again see the use of strongly saturated colours. This time the illuminated surfaces are green and blue, which will cause considerable distortion of the reflected light spectrum – probably not critical in a circulation space. In Figure 6.18 we see primary blue surfaces enhanced by the adjacency of the yellow (complementary) room surface.

It is relatively unusual to use strong saturated colours for large areas of surface in buildings. This is mainly because large areas of saturated colours demand too much attention, and most designers are aware of this. Furthermore, the effects of colour compensation in the eye, explained earlier, are quite widely understood to cause the distortion of other colours.

When unsaturated colours are used, say Munsell values of saturation value less than 3, there is sufficient white light being reflected to maintain good colour rendering. However, large surfaces of saturation 3 will give a strong colour sensation. The effect of surface size often comes as a surprise when colours have been chosen from small samples.

6.4.2 Colour temperature

This is most relevant when the designer is considering the provision of artificial light to supplement daylight,

i.e. the lighting will be in use during daytime in the darker parts of the room. If the objective is to maintain the feeling of a daylit space, then the artificial lighting should be similar in both distribution and colour to the daylight.

The colour of light sources is usually described by colour temperature. This concept has already been introduced in 6.2.1, and can be defined as the temperature of a black body that gives light of the same colour sensation as the light in question. Note that it does not have to have the same spectral distribution.

Generally the closer the match to daylight the better, but there are reservations. The main problem is that if the same sources are to be used for nighttime lighting, then it is common experience that sources of high colour temperature are unacceptably 'cold' after daylight hours. The reason for this is not obvious, but it could relate to our evolved response to the decrease in colour temperature of the setting sun, followed by the use of primitive thermal light sources such as fire and candles.

Ideally, the supplementary daylight sources would be different from the nighttime sources, although this means the duplication of lamps and circuits. Colour temperatures for common artificial light sources are given in Table 12.5.

Although the concept of colour temperature is usually applied to artificial light sources, it can be used to describe the distortion of the daylight spectrum by glass, as shown in Table 6.2. Remember that a positive shift means the light looks 'cooler', and vice versa.

Table 6.2: Colour temperature shift to daylight.

Glazing type	Colour temperature shift	(K)
Green tinted	+	400
Bronze tinted	−	740
Grey tinted	+	300
Silver coated	−	450
Low-e grey	−	1100
Low-e blue	+	2300
Triple layer polycarbonate	−	580

6.4.3 Colour rendering

Although the colour adaptation of the eye, i.e. the built in 'white balance' feature, allows us to stabilise the colour of objects under different illumination, there are limits to this. When a surface is illuminated by light with a very discontinuous spectrum, if one of the peaks in the spectrum coincides with a peak in the absorption spectrum of the dye or pigment on the surface, a disproportionate amount of energy will be absorbed. This may result in a distortion of the perceived colour when compared with the same surface viewed by daylight, with its continuous spectrum. Distortions also occur with certain strong colours when viewed under tungsten or incandescent lighting, which, although it has a continuous spectrum, is much richer in the red end of the spectrum.

This phenomenon, known as metamism, is mainly relevant in the choice of lamps for supplementary lighting or nighttime lighting. Manufacturers' lamp data will include a colour rendering index (CRI), and recommended values relating to the activities in the spaces are given in 12.4. Good colour rendering can be provided by both fluorescent sources and tungsten halogen types. It is worth noting that, to some extent, fluorescent lamps achieve high colour rendering at the expense of luminous efficacy, although even those with excellent colour rendering remain much more efficient than tungsten halogen.

In general, all modern sources have moderate to good colour rendering, and special high CRI sources are necessary only in applications such as art galleries, printing, graphic and textile studios. Another important need for good colour rendering is in medical health buildings, where diagnosis relies to a considerable extent on recognising skin colour. In some special circumstances, e.g. the display of meat for food, lighting is chosen to accentuate the colour we associate with freshness.

Finally, there has been some evidence recently to suggest that it is the continuous nature of the daylight spectrum that is conducive to health and well-being, and in artificially lit buildings this should be replaced by so-called 'full-spectrum lighting'. This can be provided, to some extent, by modern polyphosphor fluorescents with two additional rare earth phospors to simulate the red and blue ends of the daylight spectrum.

6.5 Subjective effects of colour

Most attention has been paid to the subjective effects of colour in relation to painting and the graphic arts. The perception of colour in relation to three-dimensional space has been somewhat neglected, although few architects would deny its relevance. We can identify two broad classes of effects: those effecting mood or emotion, and those related to optical effects that interact with our perception, rather than our experience, of architectural space.

The former we shall call colour associations, and the literature is full of many personal and contradictory rules. However, there is a consistent trend with which most would agree:

- Red signifies excitement, activity, danger, sex and romance.

- Orange signifies warmth, stimulation, security.

- Yellow signifies cheerfulness, happiness, movement.

- Green signifies refreshing (but) restful, calm.

- Blue signifies coolness, calm, peace.

- Purple signifies ambiguity (between red and blue), specialness, artifice, sophistication.

It is difficult to resist explaining these by reference to their occurrence in nature: the red of blood, the orange of fire, the yellow of sunlight and spring flowers, the green of vegetation and the countryside, and the blue of the clear sky and the calm sea. However, these colour associations probably represent a Western view, and are related not entirely to the natural world, but also to the cultural context. For example, in some cultures green is associated with death and decay, and is not a 'lucky' colour.

Tests have been carried out under scientific conditions but have generally failed to come up with more profound associations than those listed above. Tests in laboratory conditions have also failed to detect interaction between colour and other sensations – the often claimed feeling of warmth in rooms painted with 'warm' colours cannot be detected in objective experiments.

However, one tendency has been observed, that colours of strong saturation and higher lightness all tend to have a more stimulating effect. Thus bright apple green would have associations of freshness and excitement, where an unsaturated grey green might even be calming to the extent of being depressing.

The architectural significance of these is fairly obvious. The choice of colours in response to the function of the space is already common practice – we would not expect to see the same colours a dentist's waiting room as those in a night club or brothel.

The second class of effects we can call psycho-physical since they have partly a physical explanation. Some of these effects involve colour contrast – the effect of colour combinations. Most can be explained in terms of colour adaptation – that is, the independent desensitising of the colour stimulus receptors. First, we consider the effect of the size of the coloured surface.

Continuing the discussion concerning the impact of colour on rooms there is one more parameter beyond hue and saturation. This is size of surface, or more strictly the solid angle subtended to the eye. Small areas of colour cannot be expected to impart to the whole space the responses listed above. For example, a strong saturated red used for a filing cabinet or desk light will have an insignificant effect compared with a whole wall of saturated red. Conversely, even quite unsaturated colours and tints have a surprisingly strong impact when applied in large areas. This is often evident when viewing a floor, wall or ceiling after the colour was chosen from a very small sample.

This effect can be explained by the scanning of the eye, which constantly shifts the image over the retina and restores white balance. This actually increases the perceived colour sensation, but seems to reduce the emotional impact. This refreshing of the colour receptors becomes less possible as the field of view becomes larger – colour adaptation may even desaturate the perceived colour, but the psychological impact increases. This could lead us to propose that the perception of colour sensation is a function of relative colour, whereas the emotional impact is more to do with absolute colour. This would also be consistent with the theory that artificial

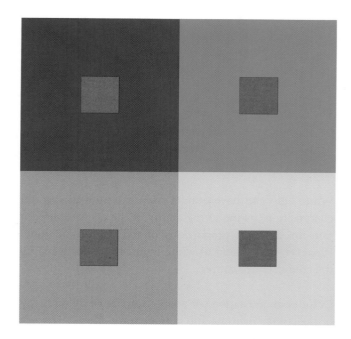

light is less 'healthy' than daylight, even if the eye percives them both as nominally white.

Colour combination is the other important area where the theory of colour perception can inform us. The colour illusion shown in Figure 6.10 and the explanation of complementarity can equally apply to the surfaces and object in a room. Figure 6.19 shows a well-known colour pattern where the neutral squares are given the chroma of the complementary colour to the surround, owing to the adaptation effect. Much the same phenomenon can be observed in rooms, where colours may appear to distort one another. Unsaturated colours are particularly vulnerable to taking on the complementary colour of a strongly coloured adjacent surface.

Figure 6.19 Complementary colours are seen in the neutral squares owing to colour adaptation. Similar effects can be created on an architectural scale.

Summary

The reflectivity of surfaces has an important effect on both externally and internally reflected light in rooms.

When light strikes an opaque surface some light is absorbed and some reflected. Reflection may be diffuse or specular; most real surfaces exhibit both to varying degree.

When light strikes a transparent layer some light is reflected, some is absorbed in passing through the layer, and some is transmitted. Transmission may be diffuse or specular. Most glasses exhibit specular transmission, allowing sharp images to be seen. Special glasses may be diffuse or a combination of specular and diffuse.

The sensation of colour is due to the spectral composition of light. Reflection and transmission at surfaces and layers can modifiiy the composition of white light sufficiently to give the sensation of colour.

The tristimulus principle is that any colour sensation can be created by a combination of three primary colour lights: red, green and blue. All three in approximately equal proportion produce white light.

White light can have quite a wide variation in spectral composition, owing to the ability of the eye to compensate for different proportions of red, green and blue. However, the 'colour' of white light can be described by the colour temperature.

Colour compensation (or adaptation) can explain many of the colour contrast and combination effects. The use of large areas of strong colours in architectural applications may distort the sensation of other colours owing to colour adaptation of the eye, or a distortion of the spectrum of reflected light, and hence must be used with caution.

References:

1. A. H. Munsell, *A Colour Notation.* The Munsell Colour Company Inc. Baltimore, MD 21218. 1967.

7 The design of windows

Window in the
library at Leuven,
Belgium.

The design of windows

The window, the transparent part of the building envelope, has always received much attention from the designer. It is usually the most visual element on the facade, and plays a major role in its composition whether it is conceived as a series of apertures piercing an otherwise opaque envelope, or a curtain wall, where the glazing takes over and becomes the wall itself (Figure 7.1).

The fact that such drastically different solutions are found side by side, and in all climates, suggests that the technical function of the window may be of rather low priority in the mind of the designer. However, the window is the most vital link with the climate, controlling the exchange of heat, light and – often – air between inside and the outside.

Windows have varied functions – frequently more than just the admittance of light. It is interesting to note that, the word 'window' in English derives from 'wind eye' referring originally to the ventilation role, and windows are still the most common elements used to provide fresh air. Windows are often the source of useful solar gains in winter, but unwelcome solar gains in summer; this conflict has to be resolved by shading.

This chapter considers the technical functioning of the window – its thermal, optical and acoustic behaviour – in relation to detailed design and materials.

a

b

Figure 7.1 The design of the glazed elements of modern buildings varies enormously and often arbitrarily, but has major implications for environmental performance:
(a) Queen's Building, Leicester, UK;
(b) CNA-SUVA Building, Basel, Switzerland.

Thermal properties

In spite of improvements in the thermal performance of glazing systems, to the extent that a double-glazed low-e argon filled window panel can have a U-value of about 1.0 W/m²K, that is nearly three times better than a solid brick wall, windows still remain the most thermally transmittive elements in most buildings. In a typical 'low energy' house design, using low-e double glazing, heat losses through the windows constitute 18% of the total conductive loss of the envelope, compared with 25% for the walls. This is because it is technically much easier to insulate opaque elements, where there is in principle no limit to the thickness of the insulation material, than to insulate the transparent element. Multiple glazing layers improve insulation, but the light transmission is diminished. Transparent insulation materials can be used, but they also diminish light transmission, and most provide only diffuse transmission, preventing a clear view.

Thus there is a compromise to be reached between maximising light transmission and minimising heat trans-mission. Table 7.1 illustrates this (a more detailed list of thermo/optical properties of glazing systems is given in 12.3). The materials are all commercially available and could be regarded as 'conventional' glazing. They are specular transmitters of light (6.1), that is the view is not distorted or de-focused.

Note that the U-value is a theoretical best value as measured in the laboratory and does not include losses through the framing components. Also, the light transmittance is at normal incidence (i.e. for light at right angles to the surface) and will be about 25% larger than that for diffuse light from all directions (6.1.2). The

ageing of the sealed units has to be considered: i.e. the slow leakage of the inert gas filling will increase the U-value.

7.1.1 Thermal comfort

As well as having an impact on heat loss, the choice of glazing material can have a direct effect on thermal comfort, particularly for those sitting close to windows. Since the closeness to a window is a positive quality in many other respects, this is obviously an important issue.

Thermal discomfort has been a common experience in the highly glazed buildings of the past three decades. In winter, this is due partly to the effect of radiation loss and partly to downdraughts, both caused by the cold inner surface of the window. In summer, discomfort is primarily from overheating by solar gain and by the heating of the person by direct radiation.

The internal surface temperature is largely controlled by the U-value of the glazing, as illustrated in Table 7.2.

In this rather severe example, apart from the comfort effects, for a typical room humidity of 50% condensation would form on the single glazing and be marginal on the double glazing. In fact frost would form on the single glazing.

The impact on thermal comfort would depend on how close the person is sitting to the window relative to its

Table 7.1: U-values and light transmittance values for various glazing types.

Glazing specification	U-value (W/m²K)	Light transmittance
Single	5.4	0.87
Double	2.8	0.75
Triple	1.9	0.65
Double low-e	1.8	0.74
Double low-e argon	1.5	0.74
Triple low-e argon	0.8	0.63

Table 7.2: Internal surface temperatures for various glazing types.

Glazing specification	U-value (W/m²K)	Internal surface temperature
Single	5.4	0.0
Double	2.8	9.6
Triple	1.9	13.0
Double low-e	1.8	13.4
Double low-e argon	1.5	14.5
Triple low-e argon	0.8	17.0

Note: Values are for room temperature = 20°C and external temperature = −10°C

area. Figure 7.2 shows that even for outdoor temperatures of –20°C, comfort is maintained almost up to the window surface for double-glazed low-e glass.[1]

Clearly there is a good case for adopting glazing of low U-value, from both an energy and a thermal comfort point of view. However, with insulation the law of diminishing returns applies; a reduction of 50% from a U-value of 5.4 (single glazing) to 2.8 (double glazing) will save more than twice as much energy as going from 2.8 to 1.5 (double low-e + argon).

Since high thermal performance glasses are progressively more costly and, as pointed out above, transmit progressively less light, we find that the choice of glazing specification will depend on the thermal climate of the site. For example, compared with single glazing, the energy saving of triple low-e + argon in Sweden would be nearly five times greater than in Portugal. Obviously this would strongly influence its cost-effectiveness.

7.1.2 Heat conduction through framing

Apart from the heat conduction through the glazing itself, heat also flows through the framing and the spacers around the edges of the multiple panes (Figure 7.3).

The thermal insulation of the framing is important, as the area of framing in elevation can be 10–25% of the opening area. Furthermore, low framing insulation values will result in low surface temperatures, leading to surface condensation on the frame.

Typical frame U-values are shown in Table 7.3, based on data given in the Norwegian Standard NS 3031. These values take account of the fact that the overall surface area of the frame is greater than the projected area of the frame – i.e. the frame acts rather as if it is a fin on a radiator.

As a general rule, it is a good idea to match the frame U-value with the glazing U-value. It can be seen that wood is the most satisfactory, combining the required mechanical properties with relatively low thermal conductivity. However, the higher mechanical strength of aluminium means that the projected area of framing can be smaller, thus reducing the heat losses through the framing compared with the glazing material. This would also reduce the obstruction to light, resulting in a smaller opening in the opaque fabric for a given daylight requirement.

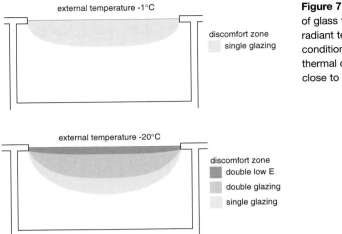

Figure 7.2 Effect of glass type on radiant temperature conditions and thermal comfort close to window.

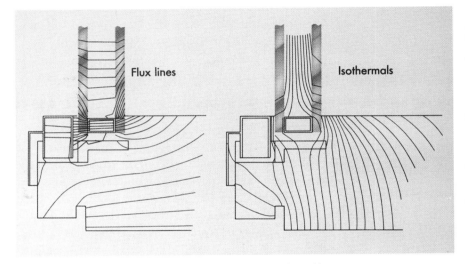

Figure 7.3 Heat flow paths through framing and spacers in multiple pane windows.[2]

Table 7.3: Typical window frame U-values.

Frame material		U-value (W/m²K)
Wood	Average thickness> 80 mm	1.6
	50 – 80 mm	2.0
	< 50 mm	2.8
Plastic	Without metal reinforcement	2.8
	With metal reinforcement	3.6
Aluminium with thermal barrier*		
Thermal path length	> 10mm	3.6
	< 10 mm	5.0
Aluminium or steel with no thermal barrier		7.0

* Thermal barriers are created by the extrusion design deliberately increasing the thermal path length by folding.

7.2 Shading design

Windows present a fundamental conflict of objectives. On the one hand they let in daylight and provide views, as well as useful solar gains in winter. On the other hand they also suffer from heat losses in winter, and glare and unwanted solar gains in summer. Glare and unwanted solar gains are usually dealt with by shading devices (sometimes referred to as solar controls), but frequently these devices also interfere with the light-admitting function.

It is all too common, on a sunny day, to see the blinds drawn and the artificial lights on (Figure 7.4). This can be experienced even in cool climates, where large areas of inappropriately oriented glazing cause glare and thermal comfort problems. This surely is a result of bad daylight design, since clearly there is no shortage of daylight, rather a problem with its control and distribution.

This section is concerned with avoiding these pitfalls, by designing shading devices that can either modulate and redistribute the intense direct sunlight, or shield the room from the parts of the sky through which the sun passes. The problem is in most cases with direct sunlight, rather than with light from the diffuse sky.

7.2.1 The objectives of shading

It is useful to set out the purpose of shading in some detail. We can identify the following:

- to minimise the total solar energy entering a room and thereby reduce the average temperature of the room

- to prevent sunlight from falling directly onto occupants, resulting in an effective increase of temperature of between 3°C and 7°C

- to reduce the local illuminance of surfaces that may present glare sources to the occupants

- to prevent the view of brightly lit outside surfaces, or clouds, or the sun itself

These primary objectives have to be met with the constraint that the daylighting conditions are not impaired sufficiently to require artificial light.

A common misunderstanding is that sunlight 'contains' heat and light, and that somehow we can separate the

Figure 7.4 Interior with blinds down and lights on.

two. In fact sunlight contains no heat – it consists of electromagnetic radiation, both in the visible region and in the invisible region (mainly infrared with some ultraviolet) in about equal proportions (Figure 7.5). It is only when radiation (visible or invisible) is absorbed by surfaces that it is converted into heat as in Figure 7.6. However, one will often see the invisible part of the spectrum erroneously referred to as 'thermal radiation', implying that only this part is the cause of heating. Obviously, for lighting, we are interested only in the visible part of the spectrum, but this still carries half the energy that potentially can become heat.

A number of parameters are used to describe the performance of shading devices. They and the relationships between them are described in detail in the glossary (Chapter 15). Here we define just two important terms, which are likely to be encountered in manufacturers' literature and codes of practice.

Total shading coefficient
This is the ratio of total transmitted radiation (including visible and invisible) passing through the window when the shading device is deployed compared with that of an unshaded, single-glazed window. The qualification 'total' indicates that it includes re-radiated energy from the shading elements or glass. Note that clear glass would have a shading coefficient of 1 and a completely opaque insulated wall a shading coefficient of zero, which is con-

trary to what one might expect. Thus a more correct term would be 'relative transmittance'.

Light transmittance

This is the fraction of visible light transmitted through the glazing and shading system if present compared with the unglazed, unshaded aperture. Both of these definitions are for normal incidence. When considering light transmittance, it is more usual to consider diffuse daylight falling onto the glazing over a range of angles. In this case the light transmittance of glass is reduced by 10% to account for the reduced transmittance at higher angles (6.1.2). The light transmittance of shading devices is highly dependent upon angle of incidence and thus cannot be described by a single factor. This will be dealt with in more detail in Chapter 9.

7.2.2 Shading types

Shading, the control of the light transmittance of the window, is commonly achieved in a number of different ways including, for example, tinted and reflective glass, retractable and adjustable louvres, fixed overhangs and lightshelves. These system have widely different physical forms and hence a wide variation in their visual impact on the building.

However, rather than classify them by their morphological characteristics, the first level of classification presented below is based on the device's daylight availability performance rather than appearance. Furthermore the performance considered is not for a single set of conditions (e.g. solar altitude and sky type), but for an integrated performance over a period of time, taking in conditions of excess illuminance and conditions of low illuminance.

It should be noted that in practice real systems do not fall precisely into these categories and may have characteristics of all three.

Three terms will be used – *retractable*, *adjustable* and *fixed*. 'Retractable' means that these elements can be completely or partially removed from the window aperture (e.g. a roller blind). 'Adjustable' means that although the element remains in the window aperture, the light transmittance characteristic can be changed, e.g. by adjusting the angle of the louvres of a venetian blind. 'Fixed' means that it has neither of the above characteristics.

Type A1: Retractable blinds, shutters and louvres

It is important to note that shading devices of this type do not influence the availability of daylight in the room in

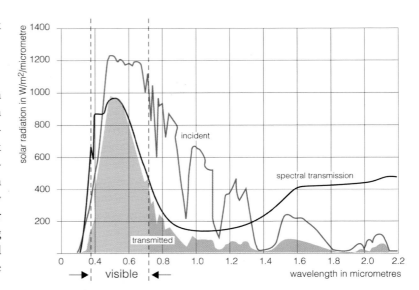

Figure 7.5 The radiation spectrum of sunlight. About half of the energy is in the visible region, the remainder in the infrared and ultraviolet. The transmission curve for clear glass is shown, indicating that some of the invisible radiation is absorbed by the glass.

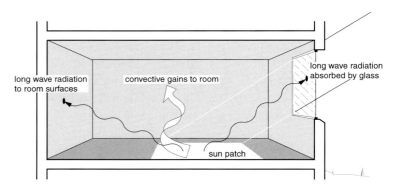

Figure 7.6 The mechanism of solar heat gain. Short-wave radiation is transmitted through the glass and absorbed by the surfaces in the room, where it is converted to heat. The glass is then opaque to the long-wave radiation from the warm surfaces. These surfaces also lose heat to the air by convection. The whole process is known as the greenhouse effect.

the limiting cases. That is, they will not influence switch-on time, because at times of low light availability they can be removed from the aperture (Figure 7.7). Clearly this property reconciles the conflict between allowing useful light in and keeping unwanted radiation out. If correctly operated, devices of this type will not lead to an increase in artificial lighting energy.

Type A2: Fixed overhangs, lightshelves and other light redistribution devices

In principle, these devices reduce the total radiation passing through the window but, owing to their being directionally selective, they do not reduce the illuminance in the darkest part of the room at times of lowest

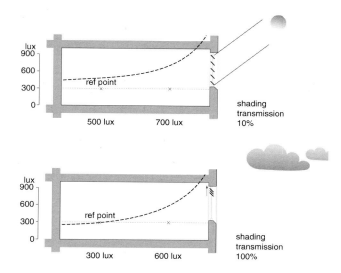

Figure 7.7 Movable shading is deployed only when there is an excess of daylight illuminance, daylight thus not affecting the limiting case when the lights have to be switched on.

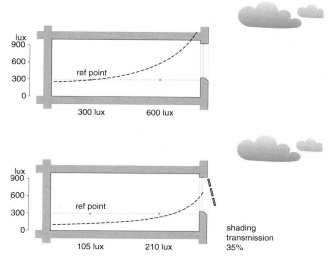

Figure 7.9 Fixed screens and reduced transmission glazing (tinted or reflecting) reduce the illuminance uniformly throughout the room and therefore increase the need for artificial lighting.

daylight availability. Thus, like Type A1, they also do not affect switch-on time and hence they do not affect lighting energy consumption (Figure 7.8).

Note that some light redistribution systems actually enhance the illuminance in the darkest part of the room. These are described in Chapter 9.

Type B: Fixed obstructing screens and reduced transmission glazing

These devices reduce all radiation – visible and non-visible, diffuse and direct – at all times in the same proportion. Thus the installation of these elements will reduce the interior illuminance at limiting times of low daylight availability and in the darkest parts of the room, and will thus affect the use of artificial light (Figure 7.9). The significance of these different types will now be described.

7.2.3 Visual and ventilation characteristics of shading devices

Daylight transmission is not the only function of the window. View out is almost always as important and, for openable windows, they also have to permit a free passage of air for natural ventilation.

Different shading devices have a different impact on these functions, but they do not necessarily fall into the same categories as their light transmission characteristics.

Figure 7.10 shows the effect of four categories of shading and their impact on vision and ventilation.

Clearly, retractable types allow unobstructed view and ventilation when retracted. However, at times of overheating risk both ventilation and shading functions are required simultaneously and thus conflict. Only in airconditioned or mechanically ventilated buildings, or in buildings with separate natural ventilation openings, would this conflict be avoided completely.

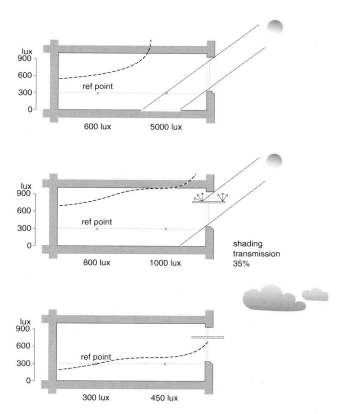

Figure 7.8 A lightshelf reduces excess illumination close to the window and rejects some radiation by reflection, but does not reduce the illumination at the critical reference point and hence does not affect when the lights are switched on.

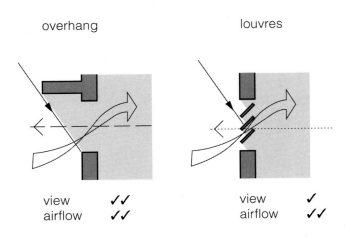

Figure 7.10 The impact of shading types on vision and ventilation.

On the other hand, louvres may have a beneficial effect on airflow if their angle is adjustable, since they can be used to control the direction of airflow to some extent. This is common practice with large 'architectural' louvre systems found in warm humid climates.

Unobstructed view is also provided with the fixed overhang and the tinted glass, and in both these cases the full shading function is maintained.

7.2.4 Retractable blinds, shutters and louvres (Type A1)

Fabric blinds and curtains

Fabric blinds can vary in reflectance and transmittance, and this has an important effect on their shading/day-lighting function. Three broad categories are illustrated in Figure 7.11.

The dark-coloured blind is a poor choice because, although it reduces transmitted radiation by a large factor, much of this intercepted energy is conveyed into the room by radiation and convection.

Light-coloured blinds have a better performance, since they reflect away a larger proportion of the visible spectrum and heat up less. However, most non-visible radiation (50% of the total) will be absorbed and most of the resulting heat will also be conveyed to the room. Unless very thick, most fabrics that appear light in colour to reflected light will also have quite high transmittance. This, together with their light-diffusing property, may well cause glare problems.

Aluminised fabrics combine high reflectance of both visible and non-visible radiation, with low transmittance. Furthermore, if the fabric is intrinsically light coloured, with the aluminium coating on the outside,

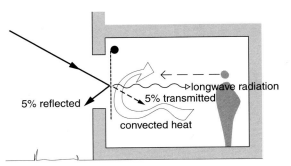

dark coloured blind: internal surface looks dark

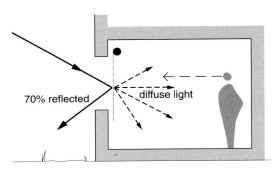

light coloured blind: internal surface looks very bright

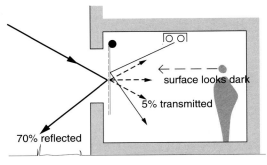

alluminised light coloured blind:
internal surface looks dark for external light but
reflective to internal artificial light

Figure 7.11 Performance of various types of blind materials.

the interior surface will look bright when viewed under artificial lighting.

Most blinds are deployed from a horizontal roller at the top of the window opening, and most curtains are parked at the side of the opening. Since this permits a continuously variable proportion of the window to be obstructed, this can give a variable transmittance to the window. However, since the unobstructed part of the window will have no diffusing or redirecting element present, a sunpatch will still exist, albeit a smaller one, and glare problems may still persist. This will often lead to the total closing of the blinds or curtains and the subsequent use of artificial lighting.

This, together with their interference with view and ventilation, constitutes a rather poor performance for occupied rooms.

To address some of these problems, blind systems employing absorptive and reflective specular transmitting films have been developed, as illustrated in Figure 7.12.

Opaque shutters

These can be considered as a kind of on/off control for daylight, allowing little or no modulation of transmittance.

Some opaque shutters may have small apertures in them, serving, in effect, the function of reducing the glazing area at times of excess radiation. However, as in the case of the partially deployed blind or curtain, the unobstructed aperture will have no light redistribution property and will still result in bright sunpatches and poor light distribution.

Louvred shutters

These combine the advantages of the retractable opaque shutter – i.e. they can be removed completely from the window at times of low daylight availability – with the light-redistributing properties of louvres when closed. When direct sunlight is present and there is the risk of overheating and glare, the louvres prevent direct sunlight penetration and allow only ground-reflected light to enter (Figure 7.13). Some light reflected from the louvre surfaces also enters, but this does not become a glare source because usually the louvres are painted with dark colours, such as green, blue or brown. Most of the ground-reflected light initially illuminates the ceiling, creating a secondary glare-free source of light within the room. This type is a very common traditional solution in Southern Europe.

Retractable and adjustable louvres – the venetian blind

This familiar form has quite a long history, originally being made of thin wooden slats (or lamellae) suspend-

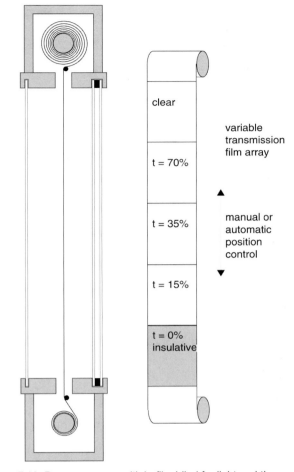

Figure 7.12 Between-pane multiple-film blind for light and thermal control.

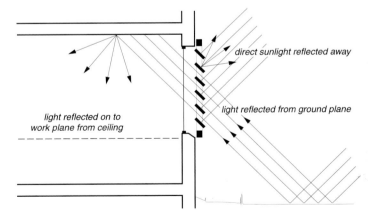

Figure 7.13 Traditional louvred shutters allow ground-reflected light to illuminate the ceiling and hence, by reflection, the work-plane.

ed on fabric webbing in such a way that the slats could be tilted through a wide angle, or raised to the top of the window out of the aperture. Modern materials such as aluminium or plastic are now more often used for the slats, and the angular control and raising and lowering are achieved with varying degrees of sophistication, including motorised control.

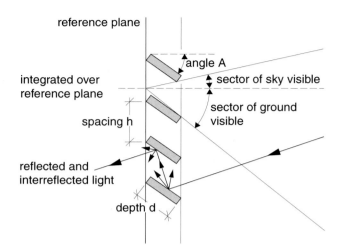

Figure 7.14 The effect of louvre angle on the diffuse light transmittance for dark louvres. See Table 7.4.

The tilting of the slats allows, in principle, an almost continuous variation in light transmittance. The slats can be adjusted until they close on themselves, where – apart from some leakage – no light is transmitted. When the louvres are at about 45° above horizontal (inside to outside), light transmittance is at a maximum. When the louvres are very thin and light in colour, this maximum transmittance is nearly as high as that of the unobstructed aperture. However, when used in the defensive mode, louvres are most often set at an angle below horizontal. The geometry is shown in Figure 7.14 and transmission factors for dark louvres are shown in Table 7.4. Note that for louvres that are light coloured light transmission is partially by interreflection between the louvres. This results in the blind becoming a diffuse light source itself as well as a source of ground-reflected light. Table 7.5 gives calculated correction factors (dif) for the diffuse transmittance factors for dark louvres.

The correction term represents the fraction of diffuse light transmitted by interreflection. The correction is itself a function of the diffuse transmission factor (DTF) since, if there were no louvres (DTF = 1), there would be no correction term. The correction is applied as follows:

$$DTF_{corrected} = DTF + (1 - DTF) \times dif$$

For thick louvres, where the thickness is more than 5% of the louvre spacing, this should be multiplied by an obstruction factor:

$$f_o = (h - t)/h,$$

where h is the louvre spacing (Figure 7.14) and t is the thickness.

Most blinds of this type are installed either inside the window panes or between panes. This is because they are rather non-rigid devices and are best protected from the wind. Indeed, this is one of their disadvantages, for although they permit the passage of air for ventilation, their light and mobile structure means that they move around in the wind, causing annoyance to some occupants.

Venetian blinds can be installed outside the glazing, but they have to be of higher specification structurally and usually require motorised operation.

In a number of systems, louvres with specular reflecting surfaces and complex cross-sections have been developed to improve the redirecting of light as well as serving the shading function. These will be described in Chapter 9.

7.2.5 Fixed overhangs and lightshelves (Type A2)

Fixed overhangs

The fixed overhang (Figure 7.15) is the simplest form of selective (Type A) shading device, relying upon the solar geometry to exclude the high angle sun, i.e. the overhang obstructs the part of the sky through which the sun passes. However, it follows that it is suitable only for south- or near south-oriented facades (in the northern hemisphere), since the solar altitude in the eastern and western sky will be too low to be obstructed by an overhang.

The ease with which south-facing windows can be shaded with overhangs, and the problems of solar control in east- and particularly west-facing windows, is sufficient to make building orientation with the long axis running east–west strongly preferred (Figure 7.16) in sites where there are no other constraints.

The obstruction of part of the sky by the overhang reduces the total amount of light flux entering the room from the diffuse sky as well as from direct sun when present. If this is not to significantly reduce the illuminance at the back of the room, there must be a strong ground-reflected component illuminating the ceiling and the underside of the overhang. This fact is often overlooked by designers – there are sound advantages in making all of these surfaces light in colour.

In this respect it can be seen that the action of the overhang is similar to that of the traditional louvred shutter, except that the underside of the overhang is above the normal viewing field and is thus less likely to create glare.

Table 7.4: Diffuse transmission factors (DTF) for dark horizontal louvres.

Ground reflectivity	0°	10°	20°	30°	40°	50°	60°	70°	80°	90°	Spacing/ depth ratio
				Angle of louvre below horizontal							
0.1	0.15	0.08	0.04	0.02	0.02	0.02	0.02	0.01	0.01	0.00	0.5
	0.28	0.21	0.16	0.12	0.09	0.06	0.05	0.03	0.02	0.00	1.0
	0.38	0.32	0.27	0.24	0.21	0.19	0.18	0.18	0.21	0.30	1.5
	0.45	0.40	0.36	0.33	0.31	0.30	0.31	0.33	0.38	0.50	2.0
0.3	0.15	0.11	0.08	0.06	0.06	0.05	0.04	0.03	0.01	0.00	0.5
	0.28	0.24	0.20	0.17	0.14	0.12	0.10	0.08	0.06	0.00	1.0
	0.38	0.34	0.31	0.28	0.26	0.25	0.24	0.24	0.25	0.30	1.5
	0.45	0.42	0.39	0.37	0.36	0.35	0.36	0.37	0.41	0.50	2.0
0.5	0.15	0.12	0.10	0.09	0.08	0.07	0.06	0.04	0.02	0.00	0.5
	0.28	0.25	0.23	0.21	0.19	0.17	0.14	0.12	0.08	0.00	1.0
	0.38	0.35	0.33	0.32	0.30	0.29	0.28	0.28	0.28	0.30	1.5
	0.45	0.43	0.41	0.40	0.39	0.39	0.39	0.41	0.44	0.50	2.0
0.7	0.15	0.14	0.12	0.11	0.10	0.09	0.07	0.05	0.03	0.00	0.5
	0.28	0.26	0.25	0.23	0.22	0.20	0.18	0.14	0.10	0.00	1.0
	0.38	0.36	0.35	0.34	0.33	0.32	0.31	0.31	0.31	0.30	1.5
	0.45	0.44	0.43	0.42	0.42	0.42	0.42	0.43	0.46	0.50	2.0
0.9	0.15	0.15	0.14	0.13	0.12	0.10	0.08	0.06	0.03	0.00	0.5
	0.28	0.27	0.27	0.26	0.24	0.22	0.20	0.17	0.11	0.00	1.0
	0.38	0.37	0.37	0.36	0.35	0.35	0.34	0.33	0.32	0.30	1.5
	0.45	0.45	0.44	0.44	0.44	0.44	0.45	0.45	0.47	0.50	2.0

Table 7.5: Diffuse interreflection factors (dif).

Louvre reflectivity	0°	10°	20°	30°	40°	50°	60°	70°	80°	90°	Spacing/ depth ratio
				Angle of louvre below horizontal							
0.1	0.32	0.31	0.29	0.26	0.22	0.17	0.10	0.03	0.00	0.00	0.5
	0.10	0.10	0.09	0.07	0.05	0.03	0.01	0.00	0.00	0.00	1.0
	0.03	0.03	0.03	0.02	0.01	0.00	0.00	0.00	0.00	0.00	1.5
	0.01	0.01	0.01	0.00	0.00	0.00	0.00	0.00	0.00	0.00	2.0
0.3	0.55	0.54	0.53	0.50	0.46	0.39	0.30	0.17	0.03	0.00	0.5
	0.30	0.29	0.28	0.25	0.21	0.15	0.09	0.03	0.00	0.00	1.0
	0.16	0.16	0.15	0.12	0.09	0.06	0.03	0.01	0.00	0.00	1.5
	0.09	0.09	0.08	0.06	0.04	0.02	0.01	0.00	0.00	0.00	2.0
0.5	0.71	0.70	0.69	0.67	0.64	0.58	0.50	0.36	0.14	0.00	0.5
	0.50	0.49	0.48	0.45	0.40	0.34	0.25	0.13	0.02	0.00	1.0
	0.35	0.35	0.33	0.30	0.26	0.20	0.13	0.05	0.00	0.00	1.5
	0.25	0.24	0.23	0.20	0.16	0.12	0.06	0.02	0.00	0.00	2.0
0.7	0.84	0.83	0.83	0.81	0.79	0.76	0.70	0.59	0.36	0.00	0.5
	0.70	0.70	0.68	0.66	0.63	0.57	0.49	0.35	0.13	0.00	1.0
	0.59	0.58	0.57	0.54	0.50	0.44	0.34	0.21	0.05	0.00	1.5
	0.49	0.48	0.47	0.44	0.39	0.33	0.24	0.12	0.02	0.00	2.0
0.9	0.95	0.95	0.95	0.94	0.93	0.92	0.90	0.86	0.74	0.00	0.5
	0.90	0.90	0.89	0.89	0.87	0.85	0.81	0.73	0.55	0.00	1.0
	0.85	0.85	0.85	0.83	0.81	0.78	0.73	0.63	0.40	0.00	1.5
	0.81	0.81	0.80	0.78	0.76	0.72	0.66	0.54	0.30	0.00	2.0

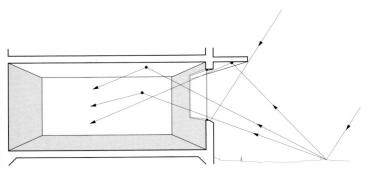

Figure 7.15 Provided there is a good reflection path via the ground and the ceiling, the simple overhang can shade the part of the sky through which the sun passes without reducing the critical illuminance level for artificial light switch-on.

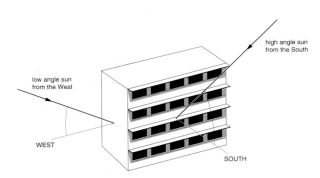

Figure 7.16 Orientation with the major facades facing north and south allows simple overhangs to be used for solar control.

In cool climates where there is the potential for solar gains to make useful contributions to winter heating, fixed overhangs are not an ideal choice. This is because the heating season lags behind the 'solar season'; there is still a significant heating load in April and May, when the solar altitude will be high and most of the window shaded. One way of overcoming this is to have a deeper overhang mounted higher on the wall. Another way is to install movable canvas awnings, as often used to protect shop windows.

Lightshelves

In deep spaces, or where the potential for ground-reflected light is poor (e.g. heavily vegetated sites), overhangs will have a significant effect on illumination at the back of the room. This would mean that they would no longer meet the Type A classification and would result in increased demand for artificial lighting.

The lightshelf offers a solution by splitting the function of the window vertically – a lower area, protected with an overhang that illuminates the front part of the room, and an upper part providing illumination for the back part of

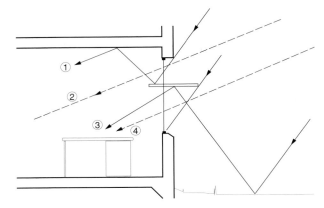

1 reflected light from shelf to back of room
2 light from diffuse sky to back of room
3 ground reflected light to front of room
4 light from diffuse sky to front of room

Figure 7.17 With a lightshelf the function of the window is split, with the lower part protected by an overhang providing light to the front of the room, and views. The upper part provides light to the back of the room, partly by reflection from the upper surface of the lightshelf.

the room (Figure 7.17). The latter is augmented by reflections from the top of the shelf (in effect replacing the ground-reflected light).

In sunny climates, conditions may be sufficiently bright to permit a further overhang on the upper window to prevent direct sunlight from reaching the back of the room. Furthermore, careful design of the geometry and optical properties of the lightshelf and glazing can result an increase in daylight penetration. This topic is dealt with in more detail in Chapter 9.

Here we are considering them only as a shading element, and we can sum up their function by stating that they are devices that eliminate direct sun and reduce the daylight factor in the front of the room, without reducing the daylight factor at the back of the room.

7.2.6 Fixed screens and reduced transmission glasses (Type B)

Fixed screens and other obstructions

In temperate climates, protecting glazing with fixed partially opaque screens achieves little since the reduction in useful daylight is the same as the reduction in unwanted radiation. In other words, apertures of smaller area but without obstructions could achieve the same performance and they would cause no obstruction to view. Furthermore, in cool climates where heat loss is of concern the opaque wall would lose much less heat than the shaded glazing.

Figure 7.18 This screen or *mashrabiya* in a house in Cairo provides light, ventilation, security and controlled visual access. This is a different set of functions from that required for conventional shading devices.[3]

The origin of this type of device, and its attraction to architects for its design possibilities, probably stem from tropical buildings where screens (often intricately ornamented) are used in unglazed apertures. Here the screen not only lets in some light, but also maintains security and privacy from without, whilst allowing ventilation (Figure 7.18). Often, in traditional buildings, indoor activities made modest claims on daylighting: shallow plans and high ceilings together with very bright skies resulted in little difficulty in daylight provision.

Clearly, this does not translate to the current objectives, which are to increase the availability of useful daylight, without creating unacceptable problems of glare and solar (thermal) gain. Thus fixed screens and other non-directionally selective obstructions are not recommended.

Solar control glasses

If it were possible to separate the invisible radiation from the visible and eliminate the former, it would go some way to achieving the objective of reducing thermal gains

Table 7.6: Typical light and total energy transmission for glazing.

Glazing type	Light transmittance	Total transmittance
6 mm float glass	0.87	0.83
Bronze absorptive	0.12	0.32
Green absorptive	0.30	0.39
Blue reflective	0.26	0.37
Green high performance	0.35	0.25
Selective high performance	0.76	0.46

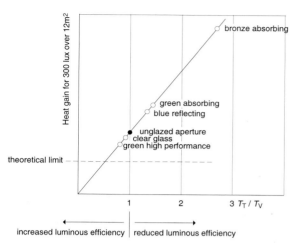

Figure 7.19 For a given level of illuminance in a room the heating effect is dependent on the ratio of the total radiation transmittance (T_t) to the light transmittance (T_v). Note that many of the older-type tinted and reflective glasses led to an increase in heat gain compared with clear glazing for a given level of illuminance.

to the room without reducing the visible light. However, in real glazing products, in absorbing or reflecting away the non-visible radiation some of the visible radiation is also lost. And we are still left with the problem that the visible light itself creates about half of the heating effect.

This problem can be illustrated by comparing the light transmittance with the total radiant energy transmittance (often referred to as shading coefficient) for a number of typical glazing products (Table 7.6). Note that light transmittance is the ratio of transmitted to incident light energy in the visible region only.

We can see that most solar control glasses reduce the light transmission by more than the total energy transmission. Only the material referred to as 'high performance' reduces the total energy transmission by more in proportion than visible radiation.

What does this mean in practice? Figure 7.19 shows that for a given quantity of light (the area is adjusted to compensate for the different light transmittivities), the heating effect is a function of the ratio of total transmittance to visible transmittance. This means that it is not advantageous to increase glazing area and reduce transmittance, unless the ratio is better than clear glass. Note that the effect of the tinted or reflective glass is to change the luminous efficacy of the light, in most cases for the worse. Also, the increased glazing area to compensate for the reduced light transmission will have implications for cost and heat loss.

Fritted glass

A recent development in solar control glass is the fritting process. This is the deposition and vitrifying of opaque ceramic dots (circular patches), usually between 1 and 10 mm diameter, onto the surface of glass, occupying between 30% and 70% of the surface. The dots are usually white, and thus reflect away a corresponding percentage of the radiation.

The mode of performance is similar to fixed screens and obstructions where both visible and total radiation is obstructed by the same amount. They also interfere with vision, although owing to the small size of the obstructions, the result is to give a hazy diffuse appearance to the view rather than the presence of a visible structure.

Since the t_T/t_V ratio is that of the clear part of the glazing, i.e. close to 1, there is no advantage to the daylight/solar control performance of the window, over clear glass of a smaller area.

7.2.7   The location of shading devices

When shading glazed apertures, some of the devices described above can be installed either inside or outside the glazing. Fabric blinds, louvres and screens are all in this category.

The light-transmitting properties will not be affected greatly, although there may be some effect due to shading of the device by the thickness of the wall – present in the case of an internal position, but possibly not present in an external position.

The greatest effect is on the thermal performance, as illustrated in Figure 7.20. With the shading device outside the glass, most of the absorbed energy is convected away to the outside. Thus the reflectance of the device does not make such a great difference, thermally. If the shading device is on the inside, most of the absorbed radiation will be delivered into the room by convection and radiation and therefore should be kept to a minimum by using a high reflectance finish. This can be extended into the near infrared (transmitted by glass) by using a metallised finish as described in 7.2.4.

The first impression is that external shading devices are greatly superior. However, they carry disadvantages. They have to be weatherproof, in particular surviving strong winds, and they often need more sophisticated mechanical control. This greatly increases their cost.

Furthermore, when windows are open for natural ventilation, the location of the shading device reduces the difference in performance.

When shading devices are used with rooflights, the penalty for an internal location is less, since convection upwards keeps the heat generated by absorption away from the occupants. This is especially true in the case of an atrium of more than one storey, when it is ventilated through the roof (Figure 7.21). In this case, internal blinds carry a very small thermal penalty, but significant reductions in cost and complexity owing to their being indoors.

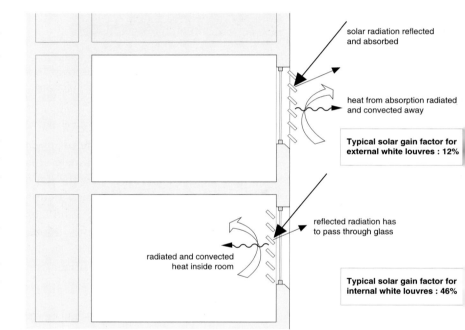

solar radiation reflected and absorbed

heat from absorption radiated and convected away

Typical solar gain factor for external white louvres : 12%

reflected radiation has to pass through glass

radiated and convected heat inside room

Typical solar gain factor for internal white louvres : 46%

Figure 7.20 The shading of windows by external and internal louvres. The difference in performance will be much greater for dark-coloured louvres, which should always be avoided if internal.

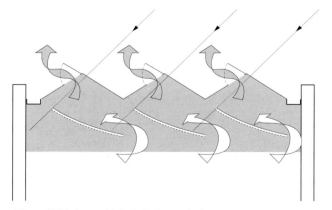

Figure 7.21 Internal blinds in the roof of an atrium. Ventilation prevents most of the absorbed radiation being conveyed to the occupants.

7.3 Acoustic performance of windows

To deal with this topic at any level of detail is beyond the scope of a book on daylight. However, the frequent use of windows to provide ventilation, and the associated problems of noise transmission, justify a brief reference.

7.3.1 Sound transmission of glazing

In most cases, the window is the weak acoustic link between the inside and outside. This is for two reasons: firstly the glass panes usually have a lower sound reduction index than the opaque wall and, secondly, when windows are opened in order to provide natural ventilation, the sound reduction becomes much smaller.

This is illustrated in Table 7.7. It shows the average sound reduction for 500 Hz, 1000 Hz and 2000 Hz. Values for this frequency band give a good indication of the reduction of sounds such as traffic noise. Please note that these values are indicative only. Specialist sources should be consulted for detailed design.

Several points are evident from this table. Firstly, a side-hung casement window when opened provides very little sound attenuation. When windows are closed, there is a small benefit in weatherstripping. Sealed double glazing makes very little difference. This is because there is no absorption in the cavity.

A major improvement is made with the double window with a large gap and reveals lined with absorbent. This demonstrates the important difference between double glazing for thermal performance and double glazing for sound reduction. A specification combining high thermal and acoustic performance would be sealed double-glazed argon-filled low-e units, with a third pane (150 – 200 mm gap) with acoustic absorption in the reveals.

Improvements can also be made in the performance of opened windows. By staggering the openings, the sound is transmitted by reflection only, increasing the chance of absorption in the reveal. Windows such as this (Figure 7.22) can achieve up to 25 dB sound reduction whilst providing significant open areas for natural ventilation.

The effect of area

As with light, the larger the window area the more acoustic energy is able to pass through it. However, because of the way we perceive sound (the dB scale is logarithmic and approximates to the sensation of loudness), doubling the amount of sound energy increases the sound level by only 3 dB. This would be the result of doubling the area of a window (in a facade where the opaque sound reduction was at least 10 dB better than the window). Thus the requirements for daylighting should never be compromised for acoustic considerations, except perhaps where extreme high noise exclusion is required for specialist purposes.

Table 7.7: Sound reduction indices for windows.

Construction		Sound reduction mid frequency (dB)
1	3 mm glazing in unsealed openable frame	20
2	As (1) weatherstripped	23
3	As (2) with sealed double glazed units	25
4	6 mm as (1)	27
5	As (1) with 2nd pane 150 mm away, reveals lined with absorbent	49
6	As (5) with 6 mm glass and 200 mm gap	51
7	Typical 225 mm plastered masonry wall	54
8	Single casement partially opened window	5–10
9	Double window as (5) with staggered openings	15–25

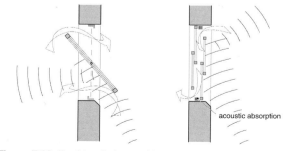

Figure 7.22 Double windows with staggered openings can provide up to 25 dB sound reduction, a significant improvement on a centre- or top-hung opening window, which acts as a reflector for traffic noise.

7.4 Summary

The improved thermal insulation value of modern low-e double-glazed argon-filled units not only saves energy but also improves thermal comfort.

With high-performance glass elements, thermal conduction through the frame has become a significant proportion of the total heat loss.

The overall objective of shading is to reduce unwanted solar gain and glare without reducing the daylight sufficiency of the room. Types of shading device that achieve this are either retractable (Type A1) or redistributive (Type A2).

Fixed screens and reduced-transmission glasses generally reduce daylight by as much as the unwanted solar gain. A smaller unshaded clear-glazed aperture will perform better in a temperate climate.

Fixed overhangs and lightshelves are redistributive devices and can shade the front part of the room without reducing the light levels at the back of the room.

Different shading devices – overhangs, louvres, translucent blinds, reduced-transmission glasses – all have significantly different impact on the view and ventilation function of the window.

External shading devices perform better than internal devices in reducing solar gain but carry penalties of cost and control. Internal devices in the roof of an atrium can work well because of their remoteness from the occupants.

The acoustic performance of windows is more influenced by the opening geometry than the glazing type. Good performance can be achieved by well separated panes with absorbing material located in the reveals.

References

1. D Button and B Pye (eds), *Glass in Buildings* (Oxford: Butterworth-Heinemann, 1993).

2. *Ibid*, p. 140.

3. H Fathy, *Natural Energy and Vernacular Architecture* (Chicago/London: Unversity of Chicago Press, 1986), p. 97.

8 Artificial lighting controls

Art Deco electric lighting at Waucquez department store, Brussels.

Artificial lighting controls

For daylight to make a real contribution to energy efficiency it is not enough that it should just be admitted into the building; appropriate lighting controls are essential. The importance of lighting control should not be underestimated. Case studies have shown that in a conventionally daylit commercial building the choice of control can make 30-40% difference to the resulting lighting use. However, in a recent study of lighting control systems in UK offices, over half the systems did not work effectively.[1]

Lighting controls cover a wide range of complexity, from manual on/off control, to centrally managed photo-responsive dimming systems including scheduling and occupancy detection. Human occupants are rather poor control detectors since although good at detecting too little light, the feedback of 'too much light', or rather 'more light than necessary', is very weak. This results in manual on/off switching being very ineffective in saving energy.

On the other hand, highly automated systems that completely remove control from the occupant are often not liked. There is growing evidence that the opportunity to exert some personal control on the local workspace environment is desirable. Systems which reach some compromise between these two extremes, including the personal control of the spatial quality of lighting by means of separating the role of task lighting and background lighting, are the most successful.

8.1 Control types

Lighting control systems have to be appropriate to the type and use pattern of the space where they are fitted. Slater *et al*[2] have identified six classes of indoor space for this purpose:

- 'Owned' spaces: small rooms for one or two people like cellular offices.

- Shared spaces: multi-occupied areas like open-plan offices, and workshops.

- Temporarily 'owned' spaces: people expect to operate the lighting controls while they are there. Examples could be meeting rooms or hotel bedrooms.

- Occasionally visited spaces: these could be store-rooms, bookstacks in libraries, aisles of warehouses or toilets.

- 'Unowned' spaces: circulation areas. People expect their way to be lit, but often do not expect to operate lighting controls.

- Managed spaces: these include atria, concourses, entrance halls, restaurants, libraries, and shops. Someone is in charge of the lighting, but usually too busy to control it. Individual users do not expect to control the lighting.

Table 8.1 suggests types of lighting control that could be appropriate to each particular space. The most effective option will depend on how much the space is occupied, and whether it is effectively daylit. An area with daylight factors below 0.5% can be classified as non-daylit. Photoelectric switching is appropriate only for areas with daylight factors above 2%, unless design illuminances are low.

Table 8.1: Recommended types of lighting control.

Space type	Daylit high occupancy	Daylit low occupancy	Non-daylit high occupancy	Non-daylit low occupancy
Owned	Manual by door Local manual Timed off manual on P/E dimming	Manual by door Local manual Timed off manual on Absence detection	Manual by door Local manual	Manual by door Local manual Absence detection
Shared	Local manual Timed off manual on P/E dimming Absence detection	Local manual Timed off manual on P/E dimming	Local manual Time switching	Local manual Absence detection
Temporarily owned	Local manual Absence detection Timed off manual on P/E dimming	Local manual Absence detection Timed off manual on Key control	Local manual Absence detection	Local manual Absence detection Timed off manual on Key control
Occasionally visited	Not applicable	Absence detection Full occupancy linking Local manual timed off manual on Key control	Not applicable	Absence detection Full occupancy linking Local manual timed off manual on Key control
Unowned	P/E dimming P/E switching	Full occupancy linking Absence detection Timed off manual on P/E dimming P/E switching	Time switching Absence detection	Full occupancy linking Absence detection Timed off manual on
Managed	P/E dimming Time switching Centralised manual P/E switching Programmed scene setting	P/E dimming Time switching Centralised manual P/E switching Programmed scene setting Full occupancy linking	Centralised manual Time switching Programmed scene setting	Centralised manual Time switching Programmed scene setting Full occupancy linking

There are five basic forms of daylight-linked lighting control:

- manual

- timed switch-off with optional manual reset

- photoelectric switching on/off

- photoelectric dimming

- occupancy detectors

These can be used in combination with each other.

8.1.1 Manual switching

Conventional manual switching has often been the default option. BRE studies[3] have discovered how occupants use switches, at least in the UK (Figure 8.1). On entering a space, people would choose whether to switch on the lighting depending on the daylight levels (Figure 8.2). If they did switch it on, they generally switched on all the lighting, and did not switch it off again until everyone had left the space.

This evidence suggests that for small spaces occupied by one or two people, such as cellular offices, ordinary manual switching will often be the best option. But in larger multi-person spaces where no one person feels responsible, wasteful use of lighting often occurs.

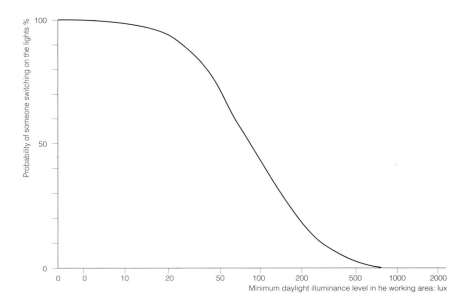

Figure 8.2 Probability that someone will switch on the lighting on entering the space (BRE).

In larger spaces with a central manual switch panel, wasteful lighting can also occur if some areas of the room are well daylit and others are not, or if some working areas are unoccupied. Localised switching is of benefit here. This can involve individual pull cords on the light fitting (Figure 8.3), switching via the building telephone system or from their personal computers – occupants can key in a code to switch the lighting on or off.

Figure 8.1
Light-switching behaviour (BRE).

9am
People switch on all (or none) of the lights when they arrive in the morning, depending on the daylight

midday
Once the lights have been switched on, they stay on all day

6pm
The lights are only switched off when the last person leaves

Figure 8.3 Pull cords allow individual control of luminaires (BRE).

Infrared switches operated by transmitter (Figure 8.4) can also be used. The transmitters can be personal, portable ones, or wall-mounted. The wall-mounted switches are battery operated, and no hard wiring is necessary.

As they need less wiring, these systems need cost no more than conventional manual switching in a new or refurbished building, yet give considerable energy savings. An additional bonus is greater flexibility if internal partitions need moving.

Localised manual switching can work effectively in daylit interiors. Sometimes daylight can brighten up a room without necessarily achieving 500 lux (or other design illuminance) on the working plane. Under these circumstances people may well choose to leave the electric lighting switched off; so there can be energy savings compared with photoelectric control. Secondly, people value individual, rapid response control. A further dimension in individual control is the localised manual dimming system (typically using an infrared controller). These are now available and are especially valued in interiors with display screen equipment.

Timed switch-off with manual reset

People are likely to switch on the lighting when they enter a room early in the morning, as daylight levels are generally low then. Later in the day, as daylight illuminances increase, the lighting will still be left on. A simple but effective solution is a timed switch-off at midday or even earlier in summer. Occupants can then choose to switch the lighting back on, but rarely do so if daylight levels are sufficient. There should not be too many switch-offs, and they should be timed to fit in with working patterns. If there are no obvious break times when most people will be out, a solar reset may be preferable.

Figure 8.4 Lighting control by individual infrared transmitters (BRE).

In this mode, the timed switch off occurs only when the daylight level exceeds a target value.

8.1.2 Photoelectric systems

Manual switching does not work so well in spaces where occupants do not feel they have 'ownership' of the light switching (unowned and managed spaces). Here photoelectric control (Figure 8.5) is a better way to link electric lighting to daylight.

In its simplest form, photoelectric on/off control switches off the lighting when a particular daylight illuminance is reached at the control point, and switches it on again when the daylight illuminance drops below this control value. This automatic switch-on is not ideal for energy saving; more sophisticated systems have manual switch-on, or switch back on only if an occupancy sensor detects movement.

Figure 8.5 Photoelectric control in Maiden Erlegh School, Reading. The outer row of luminaires is switched off and the next two dimmed, both in response to daylight (BRE).

Photoelectric switching is generally distracting to occupants, especially when the daylight illuminance fluctuates rapidly around the control value. Commercial on/off controls incorporate a time delay or illuminance differential to try to lessen this problem; the extreme approach is a 'solar reset' control that switches off photoelectrically only at a specific time of day (11.00 for example). If automatic switching on is chosen, a daylight-linked time delay (Figure 8.6) gives the best savings for a given number of switching operations. This involves not switching the lighting off until daylight has exceeded the target level for a set number of minutes. An alternative is differential switching (Figure 8.7), where switching off occurs at a different, higher illuminance than switching on. This ensures that switching off is less noticeable. Photoelectric switching is perhaps best for very well daylit perimeter areas where the lighting will be off all day anyway.

Figure 8.6 Operation of a daylight-linked time delay photoelectric control (BRE)

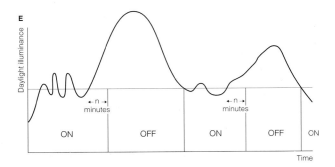

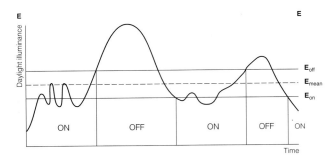

Figure 8.7 Operation of a differential switching photoelectric control (BRE).

An ideal dimming or top up control, will, like on/off control, switch the lighting off when the daylight illuminance is above a particular control value. But when the daylight illuminance is below this value this control will also increase or dim the artificial lighting so that the controlled illuminance is kept constant. Dimming controls are more expensive than on/off switching and more difficult to install, and are difficult or impossible to use with some types of lamp. However, they should save more energy, both through daylight linking and by dimming lamps at the start of their life to compensate for their increased output then. They should also be less obtrusive to occupants. For occupant satisfaction and energy reasons some form of manual reset is again desirable.

For fluorescent lighting, dimmers fall into two main types. In mains frequency control, the power consumed in the ballast is roughly proportional to the lamp output; the consumption of the cathode heaters is a constant whenever the lamp is not completely switched off. For a 1500 mm (65 W) lamp this constant component represents about 12% of the 80 W power consumption of the undimmed lamp and ballast.

Recently high-frequency dimming control has been increasingly used. This uses electronic circuitry that is around 20% more efficient than conventional ballast. Figure 8.8 shows the typical light output/power input characteristic for high-frequency controls. The lamps cannot be dimmed smoothly to total extinction. Their minimum light output is about 5% of the full value, but the corresponding power consumption is only reduced to about 25%, although it depends on dimmer type. In normal operation their residual light output and power consumption will occur throughout working hours even if the daylight illuminance exceeds the target value, unless the circuit is switched off by the occupants, an occupancy sensor or time switch.

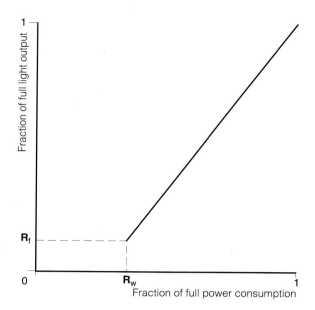

Figure 8.8 High-frequency control: graph showing energy consumed for different light outputs (BRE).

8.1.3 Occupancy detection

Occupancy sensing is not specifically daylight linked, but can give significant extra savings in intermittently occupied areas. These systems incorporate a presence detector, which in smaller spaces usually employs PIR (passive infrared) technology. For larger detection ranges microwave (Doppler shift) sensors can be more appropriate. There are two main control options: (1) full occupancy detection and (2) absence detection. Full occupancy sensing involves both presence detection and absence detection. The lighting is switched on when the sensor detects movement, and switched off when no movement has been detected for a preset period of time. This preset period varies between systems, and can in some cases be user adjustable; 5–15 minutes is common.

Full occupancy sensing with presence detection is particularly useful in spaces like warehouses where people are carrying things, driving vehicles or wearing protective clothing. It is also appropriate where people do not expect to control the lighting themselves, for example in corridors. However, in owned or shared or temporarily owned spaces presence detection can be unwelcome. People tend to resent the lighting system coming on automatically when they feel they should have control, especially when the lighting need not have been switched on anyway. Further, nuisance can occur if passers-by are detected. This wastes energy and shortens lamp life.

In these circumstances absence detection only is better. Under absence detection the occupancy sensor only switches the lighting off. Switching on is by manual control.

For both types of occupancy control, unwanted switching off, while people are present, can be a problem in some types of space. This can be because of poor-quality sensors or unsuitable sensor positioning. However, if occupants remain relatively still for some time they may not be detected since the passive infrared detectors (PIRs) rely on changes in thermal radiation caused by movement (Figure 8.9).

An occasional alternative to occupancy sensing is to use a key to switch on the lighting. This has been used to good effect in hotel bedrooms.[4] The lighting is on while a card key is kept in a special slot by the door to the room (Figure 8.10). When the room is vacated and the key taken, the lighting is switched off.

Figure 8.9 An occupancy sensor at the Queen's Building, De Montfort University, Leicester. Here occupancy sensors were installed but not used; they had trouble detecting seated occupants (BRE).

Figure 8.10 Key card switching in a hotel in Sevilla, Spain (BRE).

8.2 Daylight and energy

8.2.1 Calculation of energy saving

To demonstrate the effectiveness of controls in an existing or proposed building requires a way to calculate in advance what the lighting use is likely to be. This may be a simple manual calculation giving an annual estimate; this section describes some of the techniques. Alternatively, some computer programs can produce an hourly simulation of energy use and environmental conditions within a building.

8.2.2 Photoelectric switching

In its simplest form the on/off control switches off the lights when a particular daylight illuminance E_s (the illuminance setpoint) is reached at the control point, and switches them on again when the daylight illuminance drops below this control value. The fraction of time the lights will be off is given by C (a function of E_s), the fraction of the year for which E_s is exceeded. To get this fraction we calculate the external illuminance E that corresponds to E_s. This is given by

$$E_{out} = 100 \, E_s / (f_o \, DF) \text{ lux} \qquad (1)$$

where DF is the daylight factor (in %) at the worst-lit point in the switched zone, and f_o is the orientation factor. The orientation factor (Table 8.2) allows for the different amounts of daylight received by windows facing in different directions.[5] These factors have already been introduced in 2.1.

Lynes and Littlefair[6] give a way to estimate savings in interiors lit from one side, where the location of luminaires and control points has yet to be decided. The first step is to calculate the average daylight factor DF_{ave} in

Table 8.2: Orientation factors for an 0900–1700 day (ref 5)

	North	East	South	West
Bergen	1.04	1.27	1.60	1.15
Lund	1.01	1.19	1.66	1.23
Braunschweig	0.96	1.14	1.56	1.14
Trappes	0.96	1.19	1.52	1.11
Athens	0.97	1.26	1.68	1.25
average	1.0	1.2	1.6	1.2

the space (section 4.3.4). Then the savings are found by assuming that each half of the room is controlled separately. The average daylight factor in the front half of the room is

$$DF_{front} = 2 \, DF_{ave} \, D \, (3.056 \, H / L) / (1 + D) \qquad (2)$$

Where:

H is the height of the window head above the floor

L is the room depth measured from the window wall.

D is the ratio of DF_{back} to DF_{front}

Thus the daylight factor at the back is

$$DF_{back} = DF_{front} \, D \qquad (3)$$

and in both equations:

$$D = 1 + (1 - R_B) / (L/W + L/H + 4) \qquad (4)$$

where:

R_B is the area-weighted mean reflectance of the back half of the room

W is the window area

The values of DF_{back} and DF_{front} are then substituted into equation (1) to get the outdoor illuminance at which switching occurs in each half of the room. Then daylight availability data given in 12.2 can be used to predict the savings in each half of the room.

Photoelectric dimming

With photoelectric dimming, if the illuminance E_x on the sensor is greater than the set-point illuminance E_s then the lights will be switched off, or kept at minimum output. However, if E_x is less than the set-point the control will, in theory at least, 'top up' the illuminance to E_s (which is often called the design illuminance) by making the artificial lighting provide an extra illuminance $E_s - E_x$. In this case the fractional power saving from an ideally efficient control is given by E_x/E_s. The total hours of use are the same as for photoelectric switching but the energy use is lower because the lamps are dimmed most of the time. As a rule of thumb, the lighting energy use from a photoelectric dimming system is approximately 60% of that for photoelectric switching. This applies when a photoelectric switching system would switch the lighting off for at least 10% of annual hours. More detailed techniques are available for calculating the savings from different types of dimming control more accurately.[7]

8.2.3 Manual switching

A method has been developed for estimating the hours of lighting used by a manually switched installation. First the probable occupancy of the room is determined: in particular, the times of day when the first person enters the room and when the last person leaves. From the minimum daylight factor in the room the probability that the first person will switch on the lights can be found (Figure 8.11). The hours of lighting used are then calculated by assuming that the lights will then be left on or off until the last person leaves.

Figure 8.11 gives probability derived for Kew sky luminance data. In principle, these values could be modified for other European daylight zones by a correction factor derived from the daylight availability curve. Tentative values derived by the authors are given in Table 8.3.

Figure 8.11 also requires an orientation-corrected minimum daylight factor. Table 8.4 gives the orientation factors that are relevant here [3],[8] and which are different from the DF orientation factors. Section 4.3.4 explains how to calculate average daylight factors. Where window shape and position have not yet been decided, the minimum daylight factor can be estimated roughly by dividing the average daylight factor by 2.3.

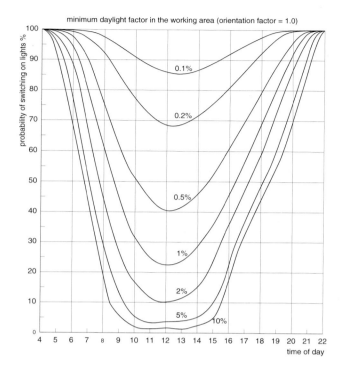

Figure 8.11 Probability of switching the lighting on, as a function of time of day and minimum daylight factor, at Kew, UK (BRE).

Various factors can affect manual switching, which the correlations above do not take into account. If the lighting buzzes, flickers or is glaring, people will want to switch it on less. The type of task is a factor; for example display screen equipment users may require less lighting. Pre-adaptation, for example before entering a room from a bright atrium or gloomy corridor, may influence switching, demonstrating that corridor luminance could influence how many people may find an office space gloomy upon entering it.

Conventional manual switching plus time switch

The energy use can be found by dividing the working day into two, before and after the time switch. Then calculate the switching probability at the start of each period, and multiply by the number of hours in the period.

Table 8.3: Correction factors for manual light switching probability.

European daylight zone	DF correction factor
1	0.78
2	1.00
3	1.13
4	1.25
5	1.38

Note: Multiply DF by correction factor before entering Figure 8.11

Table 8.4: Orientation factors derived by Hunt [3].[8]

Orientation	Factor
North	0.77
East	1.04
South	1.20
West	1.00

8.2.4 Light switching and control algorithms

A special problem with the photoelectric switch is the rapid switching of lights on and off on occasions, when daylight levels are fluctuating around the switching illuminance. This can annoy occupants and reduce lamp life. The following techniques have been developed to reduce the number of switch-offs.

1 Differential switching control (Figure 8.7). Unlike the standard photoelectric switch, the differential control has two switching illuminances: one (E_{off}) at which the lights are switched off, and another, lower, illuminance (E_{on}) at which the lights are switched on. The energy saving is close to that for an ordinary photoelectric control switching at E_{mean}, where E_{mean} is the average of the two switching illuminances E_{off} and E_{on}.

2 Photoelectric switching with time delay. Two different types of time delay are possible:

   (a) a switching-linked time delay, where switching off cannot occur until at least n minutes after the last switch on, where n is a preset delay;

   (b) a daylight-linked time delay (Figure 8.6), where switching off cannot occur until the daylight illuminance has exceeded the target value E_s for n minutes.

No delay in switching on is recommended, because that could lead to illuminances falling well below desired levels. Of the two different control strategies, a daylight-linked time delay (waiting for the daylight illuminance to exceed the target value for n minutes) gives significantly less switching. Even quite a short time delay (15 minutes or so) gives a significant reduction in switching. The number of switch-offs declines with increasing time delay, but for very long time delays (an hour or more) the returns diminish. The energy consumption for a switch with time delay will be greater than for an instantaneous on/off switch. This is because there will be times when daylight exceeds the switching illuminance but the lighting is still left on because the time delay has not elapsed.

Suppose the photoelectric control switches at an illuminance E_s and has a time delay of n minutes. The lighting energy consumption will be the same as that for a photoelectric switch without time delay, switching at a higher illuminance E_{mod}. Analysis gives the following functions for E_{mod}.

For switching-linked time delay:

$$E_{mod} = [0.07 + 0.6 \log_{10}(n + 35)] E_s \qquad (5)$$

For daylight linked time delay:

$$E_{mod} = [-0.16 + 0.83 \log_{10}(n + 25)] E_s \qquad (6)$$

For both photoelectric switching and dimming, a key decision is the location of the sensor. This will depend on the type of control algorithm used. In closed-loop control the sensor is placed in the interior and needs to allow for the output of the lighting system that it controls. This type of sensor should be screened from direct light from the windows, particularly when a daylighting system redirects light towards a ceiling-mounted photocell. Closed-loop control is inappropriate where there is a lot of task lighting; when the task lights are switched on the background lighting will dim down.

Open-loop control involves a sensor receiving daylight alone, mounted either outdoors, or indoors facing out of the window. An indoor sensor is easier to maintain and relates more closely to the daylight actually received in the space.

Sometimes a single controller dims large areas of a building. This is not a good idea since the dimming will be adjusted to provide enough light at the worst-lit point even if other areas are severely overlit. Separate control of rows of lamps parallel to the perimeter is best. Sensors at intervals along the width of the facade will be needed if some parts are more overshadowed, by buildings or trees, than others.

Sensor calibration is vital. Sometimes closed-loop controls are set up only at night; daylit calibration is also needed because of the potentially different interior luminance distribution compared with electric lighting. With open-loop control, calibration should be checked under both sunny and cloudy conditions, and periodically re-checked, particularly after redecoration or movement of internal partitions.

8.2.5 Cost-effectiveness

Lighting controls can be highly cost-effective in use, with typical payback periods from 1 to 8 years.[9] The following will generally shorten the payback time:

- Good daylighting. The techniques in the previous section can be used to assess the likely savings.

- Low occupancy. In infrequently accessed warehouses, for example, payback periods from occupancy sensing can be less than 2 years. The cost of lighting controls is often much less if they are planned as part of a new lighting installation. For example, saving the costs of wiring conventional switches can be enough to pay for a localised manual switching system.

There are other benefits of lighting control that need to be taken into account. These are:

- User satisfaction. People like to be able to adjust their lighting.

- Flexibility of space use and organisation. With some forms of localised lighting control, walls and partitions can be moved without the need to rewire wall switches. Dimming controls enable different types of task to be carried out in the same space without installing new lighting.

- Management information. Sophisticated control systems can provide feedback on luminaire use for energy management or maintenance purposes, and can identify faults.

- Reduced heat gains. Less heat comes from the lighting, an important benefit in hot climates or in buildings with high internal heat gains.

8.2.6   Integration with passive design

The form, fabric and systems of a passive solar building are arranged and integrated to maximise the benefits of ambient energy for heating, lighting and ventilation in order to reduce consumption of conventional fuels. The sensitive use of daylighting, coupled with appropriate lighting controls, can therefore be viewed as an integral part of passive solar design.

Within the EU lighting accounts for around 5% of the total primary energy consumed including that for transport and industry. However, in some types of building, such as office blocks, 30 – 60% of the primary energy (a fair reflection of energy cost) may be consumed by lighting.

In such buildings the exploitation of daylight can do much to reduce this energy cost. In case studies of buildings classified as 'energy efficient' it was found that in general the shallow plan, daylit, naturally ventilated buildings had around half the primary energy consumption per square metre of the deep-plan, air-conditioned buildings with extensive artificial lighting. Another

study by BRE indicated potential energy savings averaging 20% to 40% in offices and factories if daylighting is used effectively.[10] To achieve such savings, not only should daylight be admitted to the building but also suitable controls should be installed to ensure the displacement of energy used for electric lighting (see above).

In the design of windows daylight is only one of several factors to be considered. Windows can affect the energy balance of the building by increasing both conduction heat loss and solar gain, and to a lesser extent infiltration losses. Conduction heat loss is roughly proportional to window area. It can be reduced by using double or triple glazing, with or without low-emissivity glass. In principle such glazing causes only a small diminution in interior daylight levels (although in some 'high performance' windows the small glazing-to-frame ratio can reduce light penetration substantially). Solar heat gain is generally useful in cooler climates in winter when it reduces space heating requirements; in summer it can result in increased cooling load in air-conditioned buildings. The guidelines on the control of solar gain given in section 7.2 should be followed.

If other factors remain unchanged, an increase in the window area of a building will generally increase solar gain, and, if lighting controls are fitted, reduce artificial lighting use; but conduction heat loss will increase. The result is often that the overall energy balance in non-residential buildings does not vary greatly with glazing area [8]; there may be a shallow energy minimum at a particular glazing area, but this is hard to predict. In general, optimum window areas are higher if the windows are double or triple glazed, and if the windows are south facing. Optimum window areas will be lower for single-glazed and north-facing windows, and if suitable lighting controls are not fitted. The next section outlines the LT method, a procedure to quantify the effects of glazing area and other parameters at the design stage.

However, the type of building and its occupancy pattern also play a part. In general, because of the small variation in building energy consumption with glazing area, energy criteria are often satisfied when window areas are determined on the criteria of good daylight design.

Daylighting and solar radiation are complementary in that, when solar gains are at their highest, daylighting can be used to reduce or eliminate electric lighting, and hence the heat gains. In a passive solar building this effect is important, because it will even out swings in heat gain, reducing overheating in summer. Thus the need for good daylighting should be kept in mind at all times, especially if purely thermal elements like trombe walls are being considered. In a domestic setting, where lighting energy use is less important, the visual implications of passive solar design need to be recognised. This applies especially to north-facing rooms where minimal window areas might otherwise be chosen on thermal grounds.

8.2.7 The LT method

Amongst the considerations early in the development of a building design, the designer is concerned with two issues: the form of the building – its plan depth, section, orientation etc. – and the design of the facades; in particular the area and distribution of glazing. These parameters can all affect the energy use for lighting, heating and cooling, in a complex and interacting way. It is useful to know the implications for energy conservation of the designer's early decisions. The plan form, the facade design and the proportion of internal areas all play an important part.

The LT method (LT standing for Lighting and Thermal) is an energy design tool that has been expressly developed for this purpose.[11] A computer-based model has been used to predict annual primary energy consumption per square metre of floor area, for lighting, heating and cooling, as a function of:

- local climatic conditions
- orientation of facade
- area and type of glazing
- obstructions due to adjacent buildings (or parts of the same building)
- the inclusion of an atrium (optional)
- occupancy and vacation patterns
- lighting levels
- internal gains

The basis of the design tool is the sets of graphs (Figure 8.12), the LT curves, giving annual primary energy consumption per square metre for north, east–west and south orientations of the facade, plus one for horizontal glazing (rooflights). Curves for lighting, heating, ventilation and cooling, and total energy are generated for specific climates, using monthly average temperatures and solar radiation totals, and hourly sky luminance for a design day for each month.

The model used to derive the curves takes account of the energy flows associated with inputs for heating, cooling, ventilation and lighting, and ambient energy flows due to fabric and ventilation heat losses, solar gains and useful daylight, as illustrated in Figure 8.13.

The passive zone
The LT method uses the concept of passive and non-passive zones. Passive zones can be daylit and naturally ven-

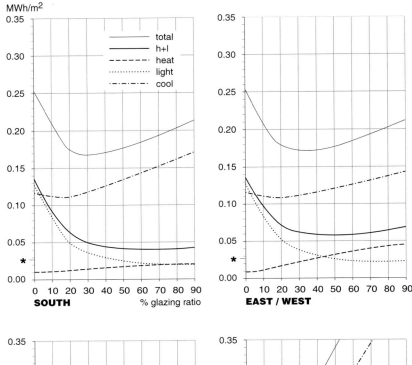

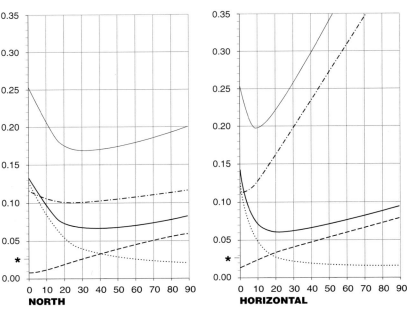

Figure 8.12 A sample set of LT curves for offices, southern UK. Lighting 300 lux, internal gains 30 W/m².

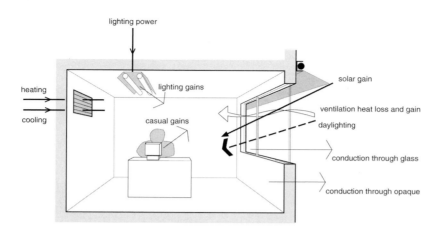

Figure 8.13 Energy flows modelled by the LT method.

tilated, and may make use of solar gains for heating in winter, but may also suffer overheating by solar gains in summer. They are defined by orientation. Non-passive zones have to be artificially lit and ventilated and in many cases cooled.

The first step in the use of the LT method is the designation of the passive zones by orientation, and non-passive zones, as in Figure 8.14. The depth of the passive zone should be limited to twice the ceiling height. All of the top floor can be a passive zone if rooflit. The zone areas are then measured off from the plan and entered into the LT worksheet, which can be either on paper or an electronic spreadsheet.

The LT curves

The vertical axis represents the annual primary energy consumption in MWh/m², and the horizontal axis is the glazing area as a percentage of total facade area. Curves are provided for vertical glazing oriented south, east–west and north, and for horizontal glazing (rooflights). Sets of curves for different building types, lighting levels and internal gain levels are available.

Two totals are shown: heating + cooling + lighting (total), and heating + lighting (h + l). The cooling curve includes an allowance for fan power as well as for refrigeration. The total without cooling can be used to indicate the energy use of a non-air-conditioned, naturally ventilated building. However, a fixed allowance for fresh air mechanical ventilation must be added for all non-passive zones. This value is given at the bottom of each set of four graphs. For buildings with high internal gains these non-passive areas would be air conditioned, and the cooling and fan power should be read off from the LT curves at the zero % glazing intercept.

The appropriate set of curves corresponding to the location and building type is selected, and a value for the glazing ratio for the particular zone is proposed. The specific energy consumption is then read off the curve and multiplied by the zone area. The result for each zone is then totalled for the building.

Procedures are available for evaluating the beneficial thermal effect of an unheated atrium or buffer

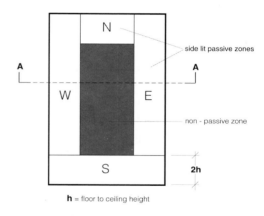

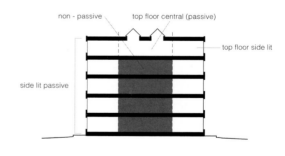

Figure 8.14 Definition of passive zones by orientation, and the non-passive zone.

space and the detrimental effect of an atrium on the availability of daylight, when compared with an unobstructed facade. There is also a procedure for calculating the effect on lighting, heating and cooling energy of overshading by adjacent buildings.

A computer-based version of the LT method applicable Europe-wide will be available during 2002. Information can be obtained on the Internet.

8.3 Summary

Providing daylight in a building does not by itself lead to energy efficiency. Even a well daylit building may have a high level of lighting energy use if the lighting controls are inappropriate. In a daylit commercial building the choice of control can make 30–40% difference to the lighting energy use.

Photoelectric control is best suited in areas like corridors, atria and entrance halls, where occupants do not expect to control the lighting. Dimming will usually give the best energy savings and will be less obtrusive. Switching is cheaper to install and may be appropriate where daylight levels are high.

In owned or shared spaces like offices, residential accommodation and some factory/workshop areas, people may prefer manual switching. A switch by the door may be enough in small spaces for one or two people. In larger areas, particularly where daylight uniformity is poor, localised manual switching is recommended. A timed switch-off, with the opportunity for occupants to switch back on manually, can give extra savings.

Occupancy sensing gives the potential for significant extra savings in intermittently occupied areas.

The LT Method offers a quick procedure for evaluating energy saving from daylight taking account of the thermal effects of glazing.

References

1. Slater A I, "Occupant use of lighting controls: a review of current practice and problems, and how to avoid them", *Proceedings CIBSE National Conference, Eastbourne, 1995* (London: CIBSE, 1995), pp. 204–209.

2. A I Slater, W T Bordass and T A Heasman, *People and lighting controls*, BRE Information paper IP6/96 (Garston, Building Research Establishment, 1996).

3. D R G Hunt, "Predicting lighting use: a method based upon observed patterns of behaviour", *Lighting Research and Technology*, vol. 12, no. 1. 1980, pp. 7–14.

4. *Energy Efficient Lighting in Buildings*, Thermie Maxibrochure (Garston: BRECSU for CEC, 1992).

5. P J Littlefair. "Predicting annual lighting use in daylit buildings", *Building and Environment*, vol. 25, no. 1, 1990, pp. 43–54.

6. J A Lynes and P J Littlefair, "Lighting energy savings from daylight: estimation at the sketch design stage", *Lighting Research and Technology*, vol. 22, no. 3, 1990, pp. 129–137.

7. *Energy Efficient Lighting in Buildings*.

8. Hunt, "Predicting lighting use".

9. *Energy Efficient Lighting in Buildings*.

10. V Crisp, P Littlefair, I Cooper and G McKennan G, *Daylighting as a Passive Solar Energy Option: An Assessment of Its Potential in Non-Domestic Buildings*, BRE Report BR129 (Garston: CRC, 1988).

11. N V Baker and K Steemers, *Energy and Environment in Architecture*, Spon Construction Series (London: Routledge, 1999).

9 Advanced daylighting systems

Roof lighting at Traphold Art Museum in Kolding, Denmark.

Advanced daylighting systems

Although windows are likely to remain the most common means of admitting daylight into buildings, they have no intrinsic properties to redirect transmitted light deep into the room. To obtain redirection effects, the openings of the building envelope need to be equipped with additional optical devices, either in addition to the windows or even incorporated into them. The resulting combinations of elements are called 'daylighting systems', frequently associated with the qualifiers 'advanced' or 'innovative', mainly to denote recent developments that incorporate new materials or products. However positive these qualifiers may seem, they do not necessarily guarantee that the systems will perform satisfactorily under all circumstances.

There are many ways of classifying daylighting systems. Here we propose three broad categories, based mainly on their geometric characteristics. As usual with this kind of classification, some systems may not fall precisely into a single category.

Reflectors and lightshelves

These are systems made of reflectors positioned either externally or internally. Since their dimensions are of the same scale as the openings they relate to, they have a large architectural impact (Figure 9.1).

Figure 9.1 The 'classical paintings' rooms at the Musée de Grenoble, France. To control daylight penetration while excluding sunlight, museum rooms are often equipped with daylighting systems comprising large reflectors suspended under zenithal openings.

Integrated window elements

These are usually made of a repetitive planar arrangement of similar tiny optical devices (e.g. miniature mirrored louvres, prismatic elements, prismatic films, transparent insulating material, laser cut panels, holographic films, etc), positioned on a parallel plane a few millimetres behind a single glass, or between two panes of a multiple glazing unit (Figure 9.2). The fact that they are integrated into windows greatly facilitates their use, especially for refurbished buildings. Apart from prismatic elements that were already in existence at the beginning of the century, the technology required to produce these systems has been developed quite recently.

Figure 9.2 The Scottish Office, Victoria Quay, Edinburgh, UK. View through a triple-glazed window with integral louvres.

Light-guides

These systems conduct natural light to the interior zones of a building, which have limited or no access to the building envelope. They consist of an outside collection device (e.g. a heliostat to capture sunlight) which redirects light into a duct with a cross-section between 0.2 m² and 2 m². As illustrated in Figure 9.3, various techniques can be used to transmit light beams through ducts. Usually these systems are designed to convey beams of sunlight, carrying large luminous fluxes through small sections.

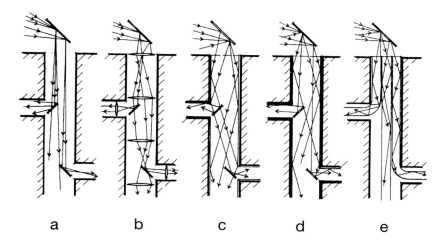

a   b   c   d   e

Figure 9.3 Five methods employed to transmit light beams through ducts deep into buildings: (a) plain duct; (b) duct with collimating lenses; (c) duct coated with a highly reflective surface; (d) duct coated with prismatic film; (e) optic fibre bundles.[1]

The main purpose of this chapter is to give criteria to assess whether a daylighting system is appropriate for a specific design and to facilitate the choice between systems with similar characteristics. After reviewing the desirable functions of daylighting systems in section 9.1, a general design procedure is proposed in section 9.2. Then the three kinds of daylighting system are discussed in more detail in 9.3 to 9.5.

9.1 The functions of a daylighting system

There are four situations where daylighting systems may be appropriate:

- Where a building is surrounded by large obstructions, systems can be designed to collect incoming light from unobstructed areas of the sky and redirect it inside the rooms.

- If the space is too deep to give adequate illuminance uniformity with conventional windows, daylighting systems can redirect parts of the incoming light flux toward the zones located furthest from the openings and thus improve uniformity.

- In sunny climates, they can serve to limit direct sunlight penetration and/or to redirect it deep into the building.

- When lighting requirements within the space are especially stringent (e.g. in museums or offices with VDUs), they can be used to reduce glare and/or to exclude sunlight penetration.

Of course it may be that some, if not all, of these issues have to be addressed simultaneously by a single daylighting system. In addition, other functions (e.g. ventilation, solar shading or provision of a clear view to the outside) may also be integrated into the same system. This multitude of requirements inevitably forces the designer to set explicit targets and priorities in order to devise a solution that represents the best compromise. The answer to the question 'What is the main purpose of the proposed daylighting system?' is of prime importance. Unfortunately, it is too often inadequately addressed from the technical point of view when daylighting systems are considered primarily for their architectural impact and aesthetic qualities. This is especially tempting with 'advanced' systems, which convey a strong 'high-tech' image.

Once the objectives are clearly defined, the way the proposed daylighting system will operate requires detailed analysis. For any daylighting system, three functions have to be fulfilled: to collect incoming light from the environment, to carry it through the building envelope and, finally, to deliver it to the interior. Light rays interact with system components in various optical ways (i.e. reflection, refraction, absorption and diffraction). These multiple interactions inevitably result in the loss of a significant part of the light flux and a modification of its spatial distribution.

In order to adapt to weather conditions or user requirements, daylighting systems sometimes incorporate movable components, either manually or automatically controlled. In these cases, the light transmission characteristics also depend on the particular position of their movable parts. For example, external louvres positioned in front of a window perform perfectly as a sun-shading device, as long as they can be correctly tilted. Meanwhile, they strongly affect diffuse light transmission both spatially and quantitatively. To ensure a good diffuse light penetration when overcast sky conditions prevail, blinds are usually designed as fully retractable systems (7.2.2).

Daylighting systems are frequently intended to serve as permanent sun-shading devices while improving the diffuse natural light penetration. A rapid analysis (Table 9.1) demonstrates that these combined require-

Table 9.1: Fraction of the visible sky hemisphere occupied by the sunpath for various latitudes, time intervals and orientations.

| Latitude | Interval | Facade orientation | | | |
		South	East–West	North	Horizon
60°N	Summer	0.13	0.28	0.04	0.26
	Winter	0.37	0.22	0.00	0.18
	Year	0.50	0.50	0.04	0.43
47°N	Summer	0.23	0.29	0.04	0.24
	Winter	0.43	0.20	0.00	0.13
	Year	0.65	0.49	0.04	0.37
38°N	Summer	0.30	0.31	0.05	0.22
	Winter	0.44	0.18	0.00	0.10
	Year	0.74	0.49	0.05	0.32

Note: The division between summer and winter is deemed to occur at the spring and autumn equinoxes. These Figures are obtained by computing the ratio of the projected solid angle of the sunpath, to that of the visible part of the hemisphere.

ments have important consequences for the design of fixed systems. With fixed shading, glazed openings in a vertical plane need to be screened from half of the visible sky hemisphere or even more, in order to reject sunlight during the whole year. For horizontal openings, at least one third of the sky hemisphere needs to be screened. This may have serious implications for daylighting.

These findings indicate that the design has to compensate for the lost part of the incoming natural light flux. When a fixed system is used in predominantly sunny climates, it has to provide a good transmission of the reflected sunlight coming from the surrounding ground plane. In a predominantly cloudy climate a fixed system is not recommended at all, since the only way to compensate the loss is to increase the area of openings by a factor at least equal to the Figures shown in Table 9.1. For example, compared with an opening with conventional glazing, a fixed system designed to totally block sunlight penetration in summer on a south facing facade needs to have an opening area at least 13% to 30% larger (depending on latitude) to provide similar performance under overcast sky conditions. This point has already been made in the classification of shading types in section 7.2.

9.2 A general design procedure for daylighting systems

A satisfactory balance between desirable and unwanted effects constitutes the real challenge when designing a daylighting system. The physical phenomena involved, the variability of the climatic conditions and the numerous parameters that can be adjusted make the problem quite complex. A methodical approach is certainly the best strategy either to help in the design of a new system or to make a choice between commercially available components. It comprises five steps that deal with the details of gradual refinements. This procedure is aimed primarily to optimise the daylighting performances for a specific building and location. It deliberately sets aside less quantifiable issues such as aesthetic quality.

The steps are as follows:

1 definition of functions and target performance

2 definition of reference case

3 detailed examination of proposed system characteristics

4 integration with the building design

5 installation and maintenance.

9.2.1 Definition of functions and target performance

As stressed above, the functions of a daylighting system must be clearly identified. In addition, target values should be defined for the quantifiable parameters that will serve to check the daylighting performance. At this stage, case studies may provide valuable help to define basic options and to set realistic target values. Since many documents presenting innovative daylighting systems originate from their inventors or manufacturers, a critical mind must be kept throughout the reading, especially if measured Figures of various performance indicators (e.g. daylight factors) are unavailable.

9.2.2 Definition of reference case

Before considering a daylighting system, at least one feasible conventional solution involving windows only should be studied first. This step is absolutely necessary,

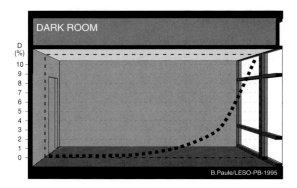

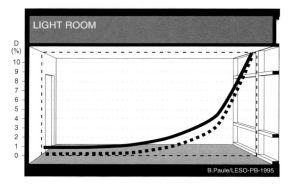

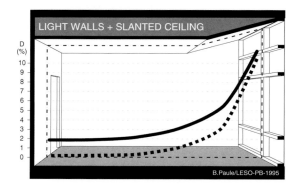

Figure 9.4 Compared with a room with very dark surfaces (reflectances: floor = 11% walls = 30% ceiling = 57%), the daylight factor profile can be very significantly raised through simple architectural measures. In the rear half of this 5.4 m deep room, the daylight factor is raised by a factor of 3.5 if the surfaces are painted with light colours (reflectances: floor = 16% walls = 75% ceiling = 75%). If, in addition, the ceiling near the facade is inclined in order to increase the window area, the daylight factor improvement reaches a factor of 7 in the rear half of the room. These values have been measured on 5.4 x 3.4 x 2.7 m test modules in the framework of a Swiss demonstration programme. A typically urban horizon has been simulated by positioning an awning-covered scaffolding in front of the facade. The height of this obstruction is about 50°.[2]

both to justify the need for a more sophisticated daylighting system from the technical point of view, and to set up a 'reference case' against which the performance of alternate designs can be objectively compared.

Simple architectural measures are also worth exploring at this level. If such measures prove to meet the desired performance, the inevitable higher costs of a daylighting system may become questionable. As an example, Figure 9.4 illustrates the improvement potential of simple architectural measures on the daylight factor distribution in the rear half of a sidelit room. This optimisation based solely on a daylight factor criterion is appropriate in a climate where overcast skies prevail.

9.2.3 Detailed examination of the proposed system characteristics

This step involves the selection or the design of the daylighting system itself. First, the acceptance of incoming light rays needs careful investigation. For this purpose some light rays incident upon the system and travelling through it should be traced on a schematic drawing (usually in a vertical section). Great attention has to be paid to ensure that the traced rays obey the physical laws governing their interactions with each component encountered. As long as the system comprises clear glazing and flat reflectors only, this task can be easily done by hand. Complications arise as soon as more sophisticated components are used (e.g. curved reflectors, prismatic components or holographic films). For this kind of system, unless ray-tracing software is used, designers have to rely on their knowledge of the light transmission or reflection characteristics of each individual component in order to draw the principal paths of the transmitted

light flux. At this level, documentation provided by manufacturers or, even better, examination of samples of the components are of great help in understanding their specific optical properties.

If sunlight penetration is intended to be controlled by the system, parallel incident rays need to be traced for several sun altitudes to ensure that they are efficiently transmitted or rejected. The orientations of the facades on which the system will be mounted, as well as the possible positions of movable elements, have to be taken into account precisely.

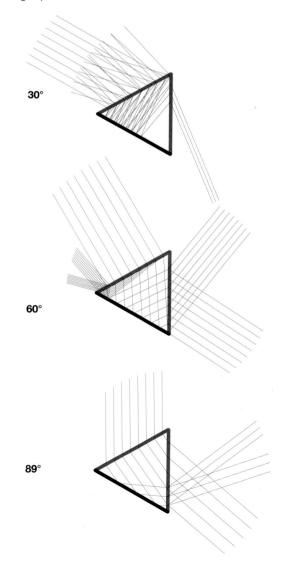

Figure 9.6 Ray paths through the triangular daylighting system as calculated by a ray-tracing program for incident beams inclined at 30° (top), 60° (middle) and 89° (bottom) over the horizon. Since the light rays encounter multiple interactions within the prismatic elements, each single incident ray may be split into more than one transmitted ray (only the two strongest transmitted rays appear in the figure).

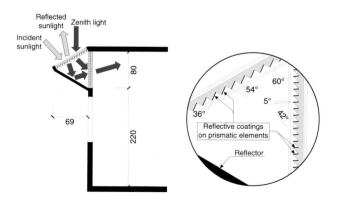

Figure 9.5 The working principle of a triangular daylighting system comprising two prismatic components and a reflector. The first prismatic element rejects sunlight while admitting light rays from high altitudes. The second prismatic element redirects the light mainly towards the ceiling of the room. A similar system has been installed on a building in Switzerland.[3]

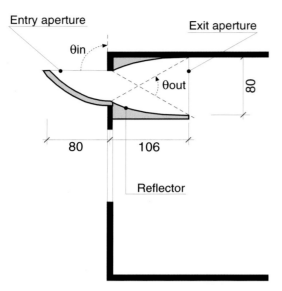

Figure 9.7 Section through the anidolic zenithal collector. The first curved reflector concentrates all incoming light from half the sky hemisphere (acceptance angle = 90°) onto a smaller vertical opening in the plane of the facade. The light rays are then shaped into a beam of well-defined angular spread (exit angle = 60°) using two curved reflectors mounted face to face.[4]

As a first example, a triangular daylighting system made of prismatic elements is depicted in Figure 9.5. As can be seen in Figure 9.6, the paths of the light rays traced through the system appear rather complex. The rejection of light rays coming from low altitudes appears to work properly. The transmission of rays from higher altitudes also works well. However, many rays appear to be delivered to the room in a downward direction. This needs further investigation, since the initial design purpose was to deliver light rays mainly towards the ceiling.

The second example is an anidolic zenithal collector. The word 'anidolic' is a synonym of 'non-imaging', constructed from two words of ancient Greek (*an* = 'without', *eidolon* = 'image'). It simply indicates that this system has been designed using methods specifically developed in the non-imaging optics theoretical framework. Originally, this theory has been applied in the field of high-energy physics. It offers a rigorous methodology for the development of daylighting systems that collect daylight and transmit it into the room with minimum losses.

As shown in Figure 9.7, this system is similar to the preceding one in that it also protrudes from the facade. Its horizontal entry aperture aims to take advantage of the higher luminance of the sky occurring at high altitudes. On its interior side, the system comprises a kind of 'daylight projector' made of two curved reflectors that

deliver the light rays to the room within well-defined angle limits ($\pm$ 30° of the horizontal plane). Figure 9.8 clearly shows that all incident rays are accepted and delivered within the desired angle. Contrary to the preceding example, this system has no inherent property to exclude sunlight since it has been designed to maximise diffuse daylight penetration under overcast skies. For this reason, an additional retractable shading device should also be mounted in front of the entry aperture.

This ray-tracing procedure proves to be a worthwhile exercise, as it forces the designer to understand clearly how the system operates and at the same time to pinpoint its limitations. It is especially useful when designing a system by assembling components whose individual optical characteristics are known but for which the characteristics of the whole system are difficult to predict by hand.

The curved shapes of the reflectors composing the anidolic zenithal collector have been designed using the non-imaging optics theoretical framework to exhibit the properties illustrated in Figure 9.8. The ray-tracing procedure is thus of lesser importance for this case, in that it serves merely to confirm the accuracy of the design.

However, this approach is limited since it relies mainly on geometrical optics and thus does not give precise indications concerning the attenuation of the light flux. Not all rays travelling through a system follow the same path. Moreover, each ray will lose a fraction of its light flux each time it interacts with a surface, either by

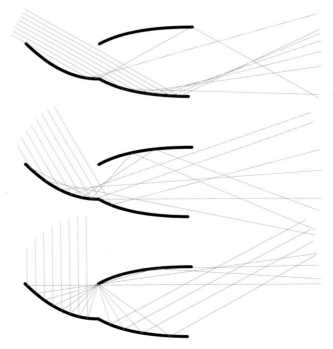

Figure 9.8 Ray paths through the anidolic zenithal collector as calculated by a ray-tracing program for incident beams inclined at 30° (top), 60° (middle) and 89° (bottom) over the horizon.

reflection from an opaque material or by absorption as it passes through a transparent layer. Consequently the transmittance depends strongly on both the position and the direction of the light rays reaching the entry aperture of the system.

For design purposes, a transmittance between 0 (ray not transmitted) and 1 (ray transmitted without any loss) should be known for each angle of incidence. The dependence of transmittance upon the exact position where each incident ray enters the system is of less interest. Therefore the directional transmittance is simply an averaged value for the whole area of the entry aperture. For an incident angle, this value represents the fraction of the incoming light flux that emerges in any direction at the exit aperture of the system. Technically speaking, this is known as a directional/hemispherical transmittance (6.1).

Since the geometry of nearly all daylighting systems or components is of a two-dimensional nature (i.e. the vertical section is extruded in the horizontal direction), it is usually sufficient to examine the directional transmittances for rays entering the system in a perpendicular plane only. This is also the same plane in which vertical sections and ray-tracing through the system are drawn (Figures 9.5 to 9.8). As shown in Figures 9.9 and 9.10, polar diagrams are appropriate to represent these values.

A transmittance between 40% and 60% is quite common for daylighting systems. However, to exclude sunlight

penetration efficiently, transmittances of 10% or even less are necessary. According to these criteria, Figure 9.9 clearly shows that the triangular daylighting system does avoid light penetration efficiently for a wide range of altitudes. On the other hand, it transmits light coming from near the zenith much less well than expected. Conversely, Figure 9.10 shows that the anidolic zenithal collector efficiently transmits light coming from the whole sky half-hemisphere.

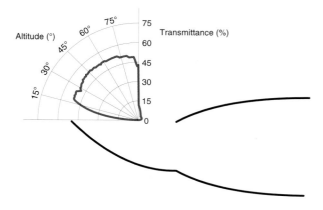

Figure 9.10 Directional-hemispherical transmittance of the anidolic zenithal collector (Figure 9.7) computed by a ray-tracing program.

By this stage, both qualitative and quantitative knowledge of how a system handles the incoming natural light flux has been acquired. This can already serve the design process, either by confirming initial choices or by providing enough solid arguments to abandon a variant.

When a commercially available daylighting system is considered, the designer should ask for the manufacturer's luminous transmittance data. Ideally, these should comprise directional transmittance Figures or, at least, give lower and upper bounds of transmittance. Unfortunately, directional transmittance data are currently rarely available, since their measurement requires a highly sophisticated measuring instrument known as a 'goniophotometer'. In the past decade, such measuring devices have been independently developed by several academic research teams.[5] Currently, five of them are in operation in Europe (in France, Germany, Netherlands, Switzerland and UK). Since the present lack of detailed data certainly constitutes an obstacle to the growth of the market for innovative daylighting systems or components, it is very likely that in the near future manufacturers will make more systematic use of these devices. At the same time, designers need to gain experience in using this kind of information.

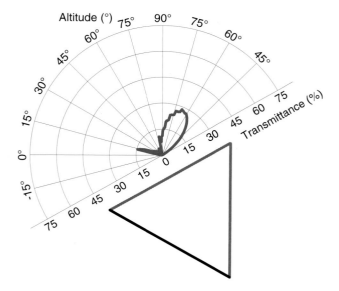

Figure 9.9 Directional-hemispherical transmittance of the triangular daylighting system (Figure 9.5) computed by a ray-tracing program.

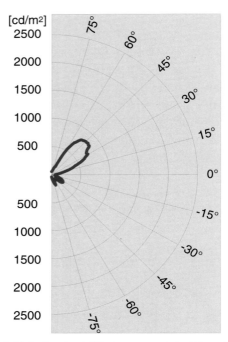

Figure 9.11 Indicatrix of diffusion computed at the exit aperture of the triangular daylighting system (10 klux CIE overcast sky with 30° external obstruction).

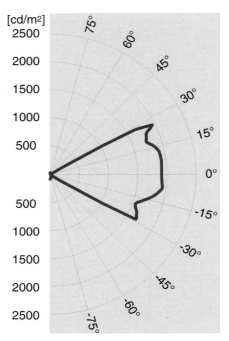

Figure 9.12 Indicatrix of diffusion computed at the exit aperture of the anidolic zenithal collector (10 klux CIE overcast sky with 30° external obstruction).

Some simulation programs are also able to compute accurate directional transmittance values for innovative daylighting systems, as long as the detailed geometry and the optical properties of the components involved are correctly modelled [6-8]. The quality of the data obtained is not only related to the capabilities of the underlying algorithms of these programs, but also depends strongly on the way they are used. Unfortunately, the learning effort and experience required to become confident in using these tools are time-consuming. Therefore computer simulation studies of daylighting systems still remain in the research sector, or in the hands of specialised engineering firms. In Europe, an ever increasing number of academic laboratories and firms are ready to offer this kind of service to designers.

The next phase looks at what happens at the exit aperture, which obviously acts as a 'daylighting luminaire' for the room. This analogy inevitably suggests drawing luminance distribution curves to describe how the transmitted light emerges from the exit aperture, providing what is termed an 'indicatrix of diffusion'. This is very similar to the technical descriptions of artificial lighting luminaires found in manufacturers' catalogues. However, owing to the variable nature of the natural light source, a daylighting system is never characterised fully by a single indicatrix. In fact a specific indicatrix is the result of both the luminance distribution of the sky and the optical properties of the system itself. To overcome this difficulty, appro-

priate assumptions should be made in order to focus on a limited number of relevant indicatrix profiles.

Bearing in mind that the two systems above are primarily intended to enhance diffuse light penetration into the room, it seems reasonable to examine their indicatrix profiles under a CIE overcast sky distribution, with an outdoor horizontal illuminance of 10 klux. To simulate the likely presence of surrounding buildings, an external obstruction extending up to an altitude of 30° is also taken into account. Figures 9.11 and 9.12 illustrate the indicatrix computed for each of our examples. To demonstrate clearly the light redirection effect provided by both these systems, an indicatrix for a normal double-glazed window is also shown in Figure 9.13.

These Figures enable further analysis of the potential performance of these systems. Obviously the triangular daylighting system will not generate any glare when viewed directly, since it delivers rays of low luminance in horizontal and downward directions (<200 cd/m²). The largest part of the light flux is directed towards the ceiling. This indicates that the ceiling should then operate a further reflection to provide light on the workplane. However, owing to its to large inclination (between 15° and 60°) and its relatively low luminance compared with a conventional window, this light beam has no real potential to increase the daylight factor in the rear part of the room.

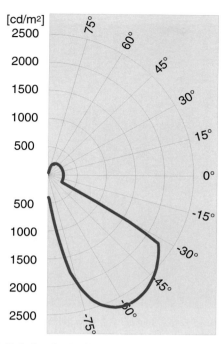

Figure 9.13 Indicatrix of diffusion computed at the inner side of a double-glazed window (10 klux CIE overcast sky with 30° external obstruction). The external obstruction significantly decreases the light flux delivered in the first 30° below the horizontal plane.

Table 9.2: Criteria for assessing potential advantages and disadvantages of high-luminance beams.

	Approximate altitude range of critical luminances
DF increase far away from the system	−30° to +15°
Glare sensation when the exit aperture is viewed directly	−50° to 0°

For the anidolic zenithal collector the findings are noticeably different. The light beam is, as expected from the design, limited to a range of directions between -30° and +30°. Its luminances are quite high (1500-2000 cd/m^2) but remain 20% to 40% below those of the sky seen through a double-glazed window. Thus both a significant increase of the daylight factor in the rear zone of the room and an improvement of the uniformity ratio can be expected. Meanwhile, some glare might be experienced when the exit aperture of the system comes into the occupants' field of view. Table 9.2 summarises the two basic criteria used to arrive at these conclusions. These criteria apply to rooms of normal geometry. Once again, indicatrices of diffusion data for existing daylighting systems are rarely available. However, they could easily be computed, at least for standard skies, from bidirectional transmittance data. As explained above, this kind of data is likely to become more common and with it the use of indicatrices of diffusion profiles. Sections 9.3 to 9.5 contain other examples of typical indicatrices of diffusion data for various daylighting systems.

9.2.4 Integration with the building design

So far, daylighting systems have been considered alone. As explained before, a careful examination of their transmittance characteristics, both at entry and at exit apertures, already appears to be of considerable help. However, even if the characteristics of a specific system seem to fit requirements, its integration in the building still needs to be addressed.

First, the design of the room needs to be checked in order to ensure that it suits the way the proposed system operates. Certain surfaces of the room are sometimes required to complement the system in order to obtain satisfactory performance. For example, a lightshelf requires the ceiling to be a secondary reflector in order to deliver daylight to the workplane. This means that the ceiling obviously needs to have a high reflectance. In addition, the ceiling should not contain cavities that act as light absorbers, such as coffers (Figure 9.14) or heavily profiled acoustic panels (Figure 9.15).

For all systems that use the ceiling as a reflector, the space beneath it should be kept free of obstructions and this necessity may interfere considerably with the desired layout of several building service systems, such as ventilation ducts, suspended luminaires, acoustic absorbing elements and sprinklers. This stresses once more the importance of the architect's considering the integration of a daylighting system at an early stage of the design process, as well as the need to notify all other consultants involved as to the implications of this choice.

In a second phase, the performance of the system installed in the building needs to be checked in order to validate the dimensioning and to evaluate the impact on energy consumption for heating, lighting and possibly cooling. Occasional under-performance or unavoidable side-effects can also be identified at this stage. This can then serve to establish recommendations aimed at future occupants regarding the control of the system and/or the layout of workplaces.

Inspection of scale models is the easiest and most common method used for this purpose. Ideally, a model should have the same photometric characteristics as the room and its associated daylighting system under investigation. The main difficulty resides in the construction of a reliable scaled version of the system. Physical simulation inevitably implies some level of approximation. In

Figure 9.14 (a) and (b): The coffered ceiling of the APU Learning Resource Centre, Chelmsford, UK, impairs the performance of the interior double lightshelf. Note that the choice of semi-transparent mirrors to redirect the light is also questionable.

a

b

Figure 9.15 UBS office building, Fribourg, Switzerland. Despite its fully glazed facade, this building offers poor daylight penetration owing to the dark colours of the surface and the presence of textured acoustic panels on the ceiling.

this context, the accurate modelling of the optical phenomena involved should be considered as a higher priority than perfectly realistic visual appearance. It is thus not absolutely necessary to use the same scale factor for modelling both the room and the components of the daylighting system. Samples of real full-scale components, such as prismatic glass or holographic films obtained from manufacturers, can sometimes be used directly in models.

Daylighting performance measurements and visual inspection of the model can then be performed by placing the model either outdoors or under an artificial sky (11.4).[9] The main advantage of these experimental facilities is their ability to model controlled and reproducible patterns of diffuse and/or direct natural illumination. Many architecture schools or faculties own such equipment. They often offer some kind of consultancy service to external professionals, or even grant access to their equipment.

Systems that are impractical to model physically can still be studied using their computer-based counterparts, handled by appropriate simulation programs. Note that although they may produce impressive pictures that look very realistic, the current rendering modules directly linked to common CAD tools do not usually have the required functions to make accurate predictions of daylighting performance. For this purpose, specialised simulation programs have to be used, which demand, as mentioned above, a certain level of expertise.[10]

Figure 9.16
Comparison of day-
light factor profiles
computed 0.8 m
above floor level in the
middle section of a
room equipped with
double glazing
(reference case), a
triangular daylighting
system and an
anidolic zenithal col-
lector. CIE overcast
sky with 30° external
obstruction.

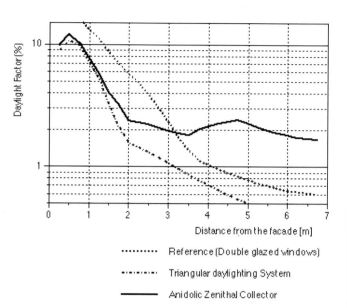

their analysis can extend beyond subjective visual inspection towards more quantitative assessment. For

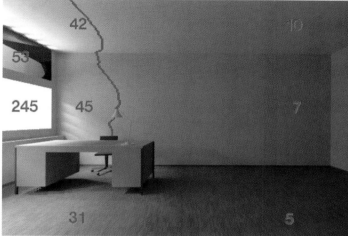

Figure 9.17 Triangular daylighting system: as already indicated by the indicatrix of diffusion, the exit aperture appears very dark.

Figure 9.16 illustrates a comparison of daylight factor profiles obtained from a ray-tracing programme. The modelled room is a rectangular office (7 x 5 x 3 m) with two facade openings (glazing ratio of 24%). The lower opening provides the visual contact with the exterior and remains double glazed for all cases. The tested systems are always mounted in the upper opening. Simulations were conducted for a CIE overcast sky with a 30° external obstruction.

As shown in Figure 9.16, the anidolic zenithal collector significantly increases the daylight factor in the rear half of the room. At the same time, it improves the uniformity ratio. Conversely, the triangular daylighting system fails to deliver daylight deep into the room. Note that both conclusions were already evident at the previous stage, when attention focused on the systems only. This confirms that, initially at least, the selection process can be based on the characteristics of the systems alone.

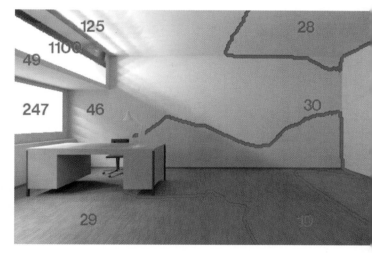

Figure 9.18 Anidolic zenithal collector: the daylight beam emerging from the exit aperture has a spectacular effect on the illumination at the back half of the room.

Computed daylight illuminance levels can be passed to thermal simulation programs to ascertain the impact of the system on the energy consumption of the building. Many additional data describing the building (e.g. materials and construction details, installed building services, patterns of occupancy, etc) and the local climate need to be collected to perform these simulations. If available at an early stage of the design process, their outcome may considerably influence the dimensioning of the relevant building services.[11]

Photorealistic pictures can also be produced by such programs. As long as a luminance is attached to each pixel,

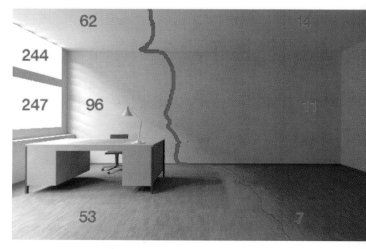

Figure 9.19 Reference case (double-glazed windows).

instance, the quantification of visual discomfort (10.1) is impossible by visual inspection alone, since the luminance range of a picture, either rendered on a VDU screen or printed on a page, is very limited compared with the range observed in the real world. In addition, over- or under-exposure in photographic reproduction can modify the appearance of a picture, while the field of luminance it represents basically remains unchanged. For our purpose, the numerical data are ultimately the most relevant. Figures 9.17 to 9.19 show computer-generated pictures on which luminance values (in cd/m^2) have been superimposed for several points in the room. To further enhance the information content of these pictures and to facilitate comparisons, 1.5% daylight factor contour lines have also been superimposed.

Daylighting simulation programs additionally offer interesting visual outputs of quantitative information.

For instance, when designing a daylit museum, the control of direct illuminance levels in exhibition areas is of major importance. Figure 9.20 shows how a false-colour picture can be used to address this kind of issue.

The construction of a full-scale mock-up positioned on a test room facade or in a real building is certainly the ultimate method to evaluate a daylighting system. However, owing to the high costs it involves, this method usually serves mainly for demonstration purposes after sufficient evidence of adequate performance has been gathered by other means.

In the framework of a Swiss demonstration programme encouraging the use of daylight in architecture, an anidolic zenithal collector was built in 1995 in a room of a modular test facility (Figure 9.21). The same facility also comprises an adjacent reference room of the same dimensions and pho-

Figure 9.20 In the first column, illuminance levels are mapped onto a false-colour scale to check how two alternative shading devices operate on the distribution of light into a daylit museum room. Top: with a system of white louvres positioned on a horizontal plane a few metres below the glazings. Middle: with a diffusing blind positioned close to the glazings. Bottom: without any shading device. The second column shows 'conventional' pictures (with pixel representing luminance levels) for the same variants.

a

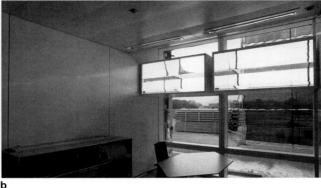

b

Figure 9.21 (a) The DIANE modular daylighting demonstration and test facility installed in Lausanne on the EPFL campus, Switzerland. The anidolic zenithal collector is mounted on the left-hand side room. (b) The anidolic zenithal collector viewed from inside.

tometric characteristics, but equipped with clear glazing only. As shown in Figure 9.22, the on-site visual comparison offers a convincing demonstration of the improvement provided by the anidolic system. Measurements of daylight factor profiles fully confirmed the performance expected from the simulations (Figure 9.23).

Figure 9.22 Fish-eye views of the reference room (a) and the room equipped with the anidolic zenithal collector (b) looking towards the back of the office room. The pictures were taken simultaneously with exactly the same settings (diaphragm and exposure time) to allow meaningful comparisons.

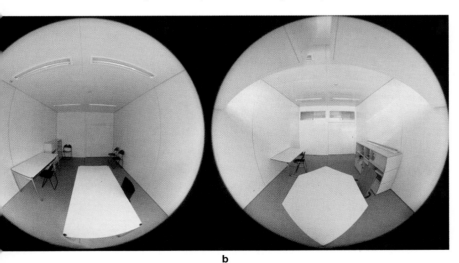

b

Figure 9.23 Daylight factor profile measured in a test module equipped with an anidolic zenithal collector. Compared with the profile measured in a dark reference room, a 10-fold increase is observed in the back half of the room (see also Figure 9.4).

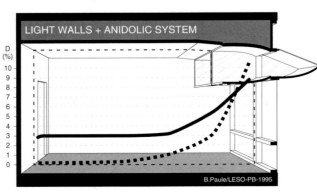

In a further phase, the possible effects of a system on surrounding buildings or nearby streets also needs attention. This is required especially for systems comprising highly reflective components visible from the outside. Reflected sunlight beams are tolerable as long as they reach critical locations during only a small fraction of the year (e.g. <10%). For particular sites (e.g. along a motorway), tougher criteria obviously apply.

Finally, construction details have to be examined carefully. For systems protruding from the facade, such as the triangular daylighting system or the anidolic zenithal collector, thermal bridges should be avoided in the structure that holds the system onto the facade. Some complications may also arise from the system itself. For example, it would be very convenient to separate the external and internal reflectors of the anidolic zenithal collector by glazing located in the plane of the facade (Figure 9.7). This would greatly simplify the construction, since the thermal barrier would remain in the same plane as the envelope. However, as shown in Figure 9.8, the rays cross this plane at angles ranging from −90° to +90°. Installing a glazed interface at this location would thus significantly reduce the performance of the system, since glass transmittance decreases strongly for angles of incidence above 60° (an overall transmittance reduction of 16% would result for this particular case). A compromise has therefore to be found between performance and ease of construction. Here it has been decided to install double glazing in the entry aperture to serve as a thermal barrier and single glazing in the exit aperture to prevent dust accumulation. This choice also means that the external reflector has to be insulated.

When the system comprises enclosed void spaces that may overheat in sunny conditions, or in which conden-

a

b

Figure 9.24 EOS office, Lausanne, Switzerland. To provide access to the external lightshelf from indoors, the apertures are equipped with small openable windows whose frames reduce the transparent area.

sation may occur during cold periods, a means of natural ventilation has to be provided. Components subject to heat will inevitably expand and hence need to be suitably fixed. Achieving watertightness may also pose some specific problems. External movable components and their fixing structures have to be designed to sustain additional loads due to wind or accumulated snow.

This incomplete list is intended as a reminder that a daylighting system needs as much attention as other more common construction details. Furthermore, as a daylighting system often becomes an essential part of the facade and hence a focus of attention, problems caused by inappropriate construction details may have a strong negative impact.

9.2.5   Installation and maintenance

Maintenance is of prime importance to ensure that the performance of the daylighting system does not deteriorate significantly. All systems inevitably collect dust and dirt that need to be removed periodically. Although rainfall has a natural cleaning effect on external glazed surfaces, this is often exaggerated. Moreover, with movable systems, mechanical failure can always be expected. Means of accessing the critical parts of the system must therefore be planned, which can sometimes

slightly impair the performance of the system. For instance, consider the lightshelf illustrated in Figure 9.24. In order to clean the reflector, it is obviously much safer and simpler to provide access from the interior than from outside. Thus the glazed apertures separating the lightshelf and the room have to be mounted in an openable frame, which reduces the transparent area.

The installation of the daylighting system in the building also requires careful attention. All components must be protected on site to avoid damage. With some components integrated in double-glazed window units (e.g. prismatic elements, miniature mirrored louvres, laser-cut panels, etc), it is easy to install them upside down or the wrong way round. Careful labelling is essential to avoid this kind of problem.

Finally, when the system is installed, time should be set aside to check that it is operating correctly. When it comprises movable parts, adjustments and periodic checks will be necessary, particularly in the first year of operation.

These considerations should further underline the need to evaluate the advantages of such a system objectively before committing the considerable effort and investment involved.

9.3 Reflectors and lightshelves

9.3.1 External reflectors

There are several examples, some even dating from early in the last century, of reflectors hung on the facade in front of windows (Figure 9.25). Their aim is generally to provide a kind of 'virtual sky' to rooms where the no-sky line remains very close to the window facade. This for instance occurs with particularly high or close surrounding obstructions.

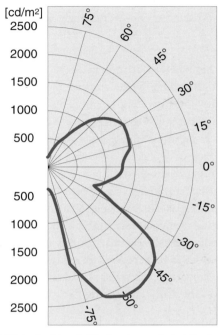

Figure 9.26 Indicatrix of diffusion computed at the inner side of a double-glazed window, in front of which an external reflector is positioned (10 klux CIE overcast sky with 30° external obstruction).

Figure 9.25 Advanced daylighting systems of the Victorian era in the City of London.

For unobstructed facades, or in sunny climates, external reflectors often combine the shading and the daylight-redirecting functions. To achieve this, the reflectors may be tilted around their horizontal axis. To improve daylight penetration during overcast periods, the reflectors have to be tilted well above the horizontal plane. Figure 9.26 shows an indicatrix of diffusion computed for a window in front of which such a tilted external reflector is positioned. By comparing this figure with Figure 9.13, the lower part of the indicatrix appears unaffected by the reflector, while its upper part shows a significant increase of the delivered light flux. This emphasises once more the essential function of the ceiling of a room.

However, when tilted at such angles, reflectors will also redirect rain towards the facade. Since this might leave dirt patterns, the maximum tilt of movable reflectors is often restricted to slightly below the horizontal plane. Their potential performance under overcast skies is thus seriously compromised.

As external reflectors are not protected from weathering, their reflectance usually does not exceed 60%. In addition, their reflection is partially diffuse, even if they are made of specular material, due to weathering.

9.3.2 Lightshelves

A lightshelf usually consists of a horizontal or near horizontal baffle positioned approximately 2 m above floor level in a facade opening. It may extend inside and/or outside the building. Two functions of the opening are thus clearly separated. Below the lightshelf, a window allows a clear view to the outside. If the lightshelf protrudes beyond the facade, this window is also shaded from high-altitude sunlight. Above the lightshelf, the reflector located on its top surface redirects daylight towards the ceiling and the rear of the room.

Since a lightshelf is a large element in the building envelope, it has a considerable impact on the architectural and structural design. The integration of the lightshelf therefore needs to be considered early on at the sketch plan stage.

Generally, lightshelves do not increase daylight factors throughout the room when compared with a conventional window of equivalent height. This is in contrast to the anidolic zenithal collector (Figure 9.7), which can be considered as an innovative type of lightshelf.

The main effect of a lightshelf is to smooth the daylight factor distribution along the depth of the room. A marked decrease is observed close to the facade, while the rear zone remains unaffected. Thus a significant improvement of the uniformity ratio is achieved, which causes the room to be perceived as relatively well lit and, at the same time, decreases the probability of occupants' switching on the artificial lighting.

In sunny conditions, a lightshelf provides shade close to the facade and, simultaneously, allows sunlight penetration to the rear zone. In winter, this helps to reduce demand for heat and lighting. However, the direct penetration of low-altitude sunlight may still generate discomfort glare for occupants. In summer, direct sunlight penetration through the lightshelf is mostly unwanted. Careful design can limit this problem, but a lightshelf should not be considered alone as an efficient shading device. With the south-facing lightshelf illustrated in Figure 9.24, the clerestory window is located slightly behind the plane of the facade. It is thus permanently shaded from direct high-altitude sunlight. All sunlight reaching the window arrives having lost at least part of its energy through reflection.

Tilted elements also provide interesting solutions. In Figure 9.27, a tilted south-facing lightshelf at below sill level and its associated tilted glazing provide seasonal selection. In winter the sun rays are reflected towards the ceiling of the room, whereas in summer they are mostly rejected. In this unusual configuration, the lightshelf is specular and is located outside just below sill level.

The control of diffuse light penetration can also be adjusted through design. Analysis of the light rays transmitted by a horizontal lightshelf shows that those rays coming from the higher unobstructed parts of the sky reach the

Figure 9.27 Collège la Vanoise, Modane, France. A tilted external lightshelf oriented towards the south acting as a sunlight seasonal control device. Unusually, the lightshelf is below sill level.

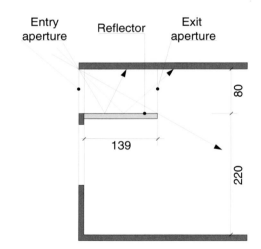

Figure 9.28 Section of a horizontal internal lightshelf.

ceiling very close to the facade (Figure 9.28). Thus they contribute only marginally to illuminating the rear of the room. Ideally, a lightshelf should redirect all incoming rays further into the room. This can be achieved by an anidolic reflector (Figure 9.29).[12] It appears that the curvature of the reflector requires the entry aperture to be resized in order to keep the dimensions of the whole system similar to those of a horizontal lightshelf.

A comparison of their indicatrices of diffusion profiles indicates that both systems provide roughly identical light flux at their exit aperture (Figures 9.30 and 9.31). However, the anidolic version has greater potential to direct the light deeper into the room, since it distributes

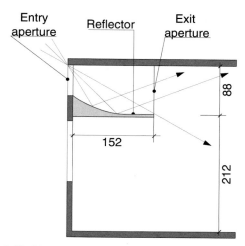

Figure 9.29 Section of an anidolic internal lightshelf.

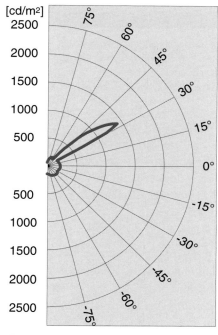

Figure 9.30 The indicatrix of diffusion at the exit aperture of a horizontal internal lightshelf.

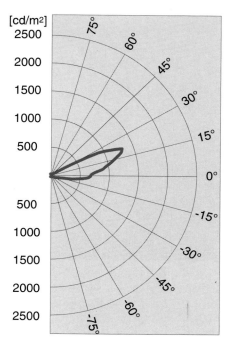

Figure 9.31 The indicatrix of diffusion at the exit aperture of an anidiolic lightshelf.

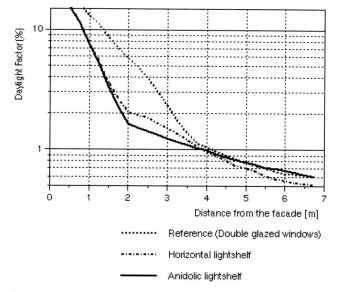

Figure 9.32 Comparison of daylight factor profiles computed 0.8 m above floor level in the middle section of a room equipped with double glazing (reference case), an horizontal lightshelf and an anidolic lightshelf (CIE overcast sky with 30° external obstruction).

its light flux at altitudes between −10° and 30°, while the lightshelf has a marked peak above 30°.

As shown in Figure 9.32, simulations have confirmed that, despite the reduction of its entry aperture size by 1/3, the anidolic lightshelf provides slightly higher daylight factors at the back of the room than the horizontal lightshelf. This means that illuminance levels can be improved, while the energy transfers through the clerestory window (solar gains and thermal losses) are reduced by a third. This interesting result is strictly equivalent to an increase of the luminous efficacy of daylight by the same factor. This example demonstrates that the intrinsic properties of a lightshelf can be improved by careful design.

Other parameters may also be varied to achieve desirable properties. For instance, specular reflectors can be replaced by semi or totally diffusing materials, with the aim of smoothing the image of the sunpatch projected onto the ceiling. However, they are less efficient than specular reflectors for enhancing diffuse light penetration and are thus recommended mainly for sunny climates.

Although lightshelves have already been used for several decades, they are still generating novel design solutions.

Figure 9.33 The dark-coloured ceiling (reflectance = 28%) of the Agricultural Bank in Athens, Greece, cancels the daylight redistribution effect of the lightshelf. In this south-east oriented room, the mean daylight factor amounts to 0.58% only.

Figure 9.34 Queen's Building, Leicester, UK. Great attention has been devoted to the clerestory part of the lightshelf in order to achieve efficient daylight penetration; however, occupants complain about the limited view from the windows.

This results from the great freedom that remains in their design, since a large number of geometric and construction parameters can be adjusted to suit different applications. Lightshelves are not standard products, and their design is really part of the architectural process. However, the great impact of a lightshelf on the internal and/or external appearance of the building facade sometimes tempts the designer to neglect the function of the system as a daylight provider. For instance, the ceiling of the room, which does not belong to the system itself, plays an important role as a secondary reflector. To work effectively, therefore, a lightshelf requires a relatively high ceiling of light colour, without protruding or hanged elements that could intercept light rays. As already shown in Figure 9.14 and also in Figure 9.33, these essential requirements are not always satisfied.

On the other hand, if the design focuses mainly on the maximisation of daylight penetration, there is a risk of forgetting other requirements that have great influence on occupant satisfaction. For example, the large internal lightshelf illustrated in Figure 9.34 provides high and uniform daylight factors throughout the room. However, nearly half the occupants complain about the small size of the viewing windows.

9.3.3 Anidolic zenithal openings

Anidolic reflectors also offer interesting features when applied to the design of daylighting systems for roof openings. To optimise the daylight penetration while excluding sunlight penetration, the entry aperture has to be tilted to the north. Taking advantage of its sharp angu-

lar selection property, the first anidolic collector is positioned so that its admission sector includes the whole sky between the horizon and the highest position of the sun. For a specific location, the highest altitude of the sun (occuring at summer solstice) can be calculated by

$$\text{highest altitude} = 90° - \text{latitude} + 23.45°$$

The sun never appears inside the admission sector, except at the beginning and end of days between the spring and autumn equinoxes (Figure 9.35). Thus the shading is completed with vertical-shaped slats laid uniformly over the aperture; the spacing between them is fixed at 0.5 m. As shown in Figure 9.36, the collected light flux is then directed downwards by a connecting element. Finally, two symmetrical reflectors disperse the output beam.

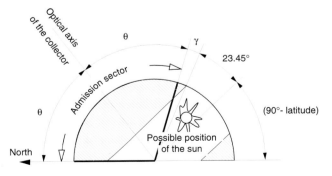

Figure 9.35 Admission sector of an anidolic zenithal opening. A small tolerance angle γ can be introduced to also reject light coming from the brighter portions of the sky near the sun direction around summer solstice.

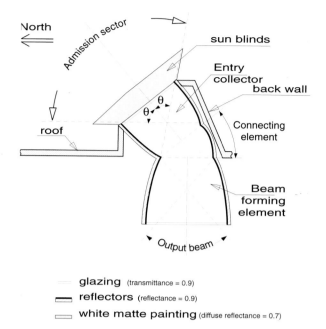

Figure 9.36 Cross-section of an anidolic zenithal opening.

Although not of the same geometry, commercially available reflective microlouvres assembled in an eggcrate structure are advertised for the same purpose. They can be installed on a plane a few centimetres thick hanging below a glazed roof. Unlike the anidolic zenithal openings, these louvres can easily be fitted into existing buildings. This can, for instance, serve to counteract over-heating or glare problems encountered when zenithal glazed openings have been oversized.

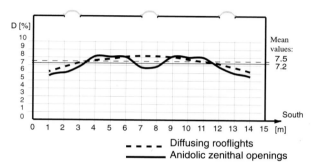

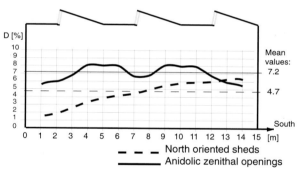

This kind of system has not yet been built in a real building. However, detailed simulation studies indicate that its potential performance is very encouraging when compared with two conventional systems employed on zenithal openings, i.e. diffusing rooflights and north-facing sawtooths.[3] Figure 9.37 shows a comparison of daylight factor profiles. Although the anidolic zenithal openings exclude a large part of the sky hemisphere to avoid sunlight penetration, they provide almost the same daylight factor levels as diffusing rooflights of the same opening area. Compared with north-facing sawtooths, the anidolic zenithal openings significantly increase daylight factor levels.

Since the beams of daylight emerging from the exit apertures are divergent, anidolic zenithal openings also perform favourably for the control of discomfort glare. Thus this kind of system should find favourable application for indoor spaces for which visual comfort is essential, e.g. sport halls, museums, and retail spaces.

Figure 9.37 Comparisons of daylight factor profiles. The simulated building is typical of a factory hall (dimensions: 15x10x7 m; reflectances: floor = 0.3, walls = 0.5, ceiling = 0.7). Daylight is provided by three rectangular top openings, aligned to the east–west axis. The opening aperture area is equal to 20% of the roof area. The diffusing rooflight's characteristics are typical of commercially available products (directional transmittance = 0.06; diffuse transmittance = 0.56). The sawtooths have their openings facing north and are equipped with clear double glazing (transmittance = 0.81).

9.4 Integrated window elements

9.4.1 Prismatic elements

Prismatic elements have been used in buildings for over 50 years, mainly to redirect diffuse sky light. They are currently incorporated in innovative daylighting systems, where they also serve to redirect sunlight.

Prismatic elements are made of transparent materials (usually transparent polymers) shaped as planar elements with a flat face on one side and prismatic facets positioned in a regular pattern on the other side. They differ mainly by the angles separating the facets. Various types are available on the market, either as square panels approximately 25 cm in size and 1 cm thick, or as thin flexible films less than 1 mm thick.

Prismatic elements take advantage of two physical phenomena: refraction and total internal reflection. Since rays that travel inside prismatic elements may undergo either of these two interactions several times before leaving, an accurate prediction of their directional optical properties is difficult to obtain without a dedicated computer program. Each type of prismatic element has principal characteristics that are suited for either ray rejection or redirection.

For instance, Figure 9.38 illustrates a prismatic element where facets make an angle of 90°. Owing to this geometry, light rays incident on the flat side with angles between ±5° around the normal plane undergo total internal reflections at each facet and hence are finally rejected backward, as indicated by the zero transmittance. Within this narrow angular range, the prismatic element behaves like a perfect mirror. For all other incident angles, light rays are transmitted well although their paths are altered.

Prismatic shading devices can be designed to take advantage of this property. Figure 9.39 shows such a system, as viewed from inside. The upper part of the opening is made up of a double-glazed unit inside which a prismatic element of the same type as depicted in Figure 9.38 is encapsulated. The facets of the prismatic element are horizontally positioned, and the whole system can tilt around on a horizontal axis. This enables the system to be adjusted so that the sun direction remains exactly in the normal plane in which all incident rays are rejected backward. Under an overcast sky, the system still transmits a significant amount of diffuse light towards the interior.

This kind of system sometimes generates glare, since the brightest part of the sky located close to the sun direction may extend beyond the narrow angular zone for which rays are rejected. Thus additional blinds or other shading devices may still be needed.

As shown in Figures 9.39 and 9.40, the prismatic elements prevent a totally clear view through. This problem can be solved by installing complementary prismatic elements at a very close distance from the first elements. However, owing to the presence of an additional transparent layer, a slight reduction of the overall transmittance would result.

Prismatic elements can also be manufactured with specific facets made reflective by an aluminium coating. Since this treatment greatly enhances their angular selectivity, these elements are usually incorporated in fixed systems that prevent sunlight penetration either for a specific period or even for the whole year. Figure 9.5 illustrates such a system.

Owing to the large part of the sky that these systems permanently shade, they are unlikely to provide high

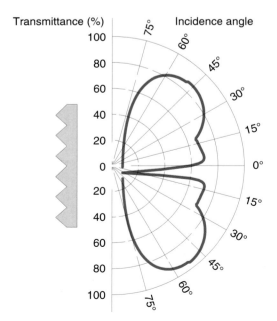

Figure 9.38 Directional-hemispherical transmittance of a prismatic element.

Figure 9.39 A prismatic system installed in the upper part of a double-skin facade of the CNA-SUVA office building in Basel, Switzerland. The system is automatically tilted in order to avoid sunlight penetration.

Figure 9.40 An external view of the facade of the CNA-SUVA building. The prismatic shading systems prevent the clear view of the window area in front of which they are installed.

daylight factors. Thus they are not recommended for climates with frequent overcast sky conditions.

Apart from the sunlight rejection function, prismatic elements can also serve to redirect the incoming light flux in order to enhance sunlight and/or diffuse light penetration. For this purpose they can be mounted on vertical windows. In sunny conditions, sunlight is redirected towards the ceiling, which significantly improves the illuminance levels in the adjacent room. Unfortunately, unwanted downward light beams are often produced at the same time, and these may cause glare problems.

In overcast conditions, measurements taken in mock-up test rooms have shown that this kind of system decreases daylight factors slightly compared with a similar room with clear glazing. However, simulation studies have shown that, in highly obstructed locations, prismatic systems can redirect skylight deeper into the room and therefore significantly improve daylight factors. These findings indicate that the performance of redirecting prismatic elements strongly depends on detailed design and site characteristics (i.e. orientation of the facade, surrounding obstructions, etc). Modern simulation tools are required to take these parameters into account.

Prismatic elements are not limited to sidelighting. For instance, a recent system has been developed for rooflights. It comprises two parallel discs made of prismatic elements. By rotating each disc separately, the system automatically adjusts its optical properties in order to catch sunlight and deliver it in a downward direction. In addition, the system can operate in a dimming mode, in which the positions of the discs are automatically controlled to provide a relatively constant illuminance level.

This system is now entering the market and has already been installed in a few buildings in Japan, where satisfactory performance has been recorded. However, it is again a system that works mainly with sunlight, and its price is currently high.[14]

9.4.2 Transparent insulating materials

A large variety of so-called 'transparent insulating materials' (TIM) are now available on the market (e.g. honeycomb structures, aerogel, etc). Mounted in the gap that separates the two panes of a double glazing unit, they significantly decrease the U-value and still allow good transmission of light (typical transmittance between 50% and

60%).

When installed in front of a wall, this kind of system forms a well-insulated translucent layer that greatly affects the heat balance: over a year the wall behaves as a passive solar collector. This technique is particularly interesting for building refurbishments. However, in climates with warm or hot summer periods, shading devices to prevent overheating are essential, but add significantly to the cost.

Transparent insulating materials can also deliver daylight to interior spaces (avoiding high heat losses), for which purpose they are installed in windows.

As shown in Figure 9.41, the indicatrix of diffusion for a window equipped with a typical transparent insulating material is nearly uniform. Viewed from inside, the window appears as a large screen of constant luminance. No light is beamed in a particular direction. Such a system thus does not increase daylight factors in the room, unless it is installed on a site surrounded by high external obstructions. However, the uniformity ratio is improved and the luminance contrasts between the windows and the surrounding areas are reduced. This contrast grading effect (10.1.8) contributes to the limitation of discomfort glare under overcast sky conditions. Conversely, when sunlight reaches these systems, they become large sources of glare in the field of view. As shown in Figure 9.41, the diffusion of sunlight may result in worse conditions, since

the surfaces that would be directly seen through the same openings have often much lower luminances. It is thus appropriate to consider transparent insulating materials a last resort method for controlling glare.

Note that the name given to these materials is particularly misleading, since all of them are translucent, not transparent. This implies of course that they do not provide a clear view. For certain applications this can be considered an advantage, either to preserve privacy or to hide unpleasant or uninteresting surroundings (Figure 9.42).

Compared with older materials, such as glass bricks (Figure 9.43) or translucent glass (Figure 9.44), transparent insulating materials perform very similar optical functions. However, they offer much better thermal insulation properties.

9.4.3 Other types of integrated window element

Owing to their ease of installation and maintenance, integrated window elements are often considered to be

Figure 9.42 Super-insulated glass house, Ballerup, Denmark. This totally glazed experimental house uses advanced glazing and transparent insulating materials to reduce heat loss and to collect solar radiation. Translucent walls made of this latter material maintain some level of privacy.

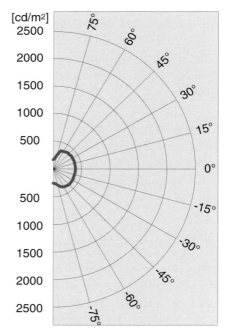

Figure 9.41 The indicatrix of diffusion of transparent insulating material.

Figure 9.43 Domestic multi-storey building, Fribourg, Switzerland. Glass bricks installed in a living room wall. The resulting diffusing effect 'opens' the room while preserving privacy. This could also have been achieved with transparent insulating materials, but would lack the patterned effect of the joints between the glass bricks.

Figure 9.44 Viewed from outside (a) during daytime, translucent elements (here translucent glazing) give the impression of an opaque envelope. (b) Viewed from inside the wall becomes luminous. Stansted Airport, UK.

a

b

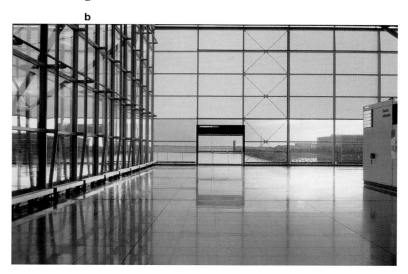

the most promising type of daylighting system. Thus several innovative elements have been developed of this kind since the 1980s and are currently beginning to penetrate the market.[15, 16] We briefly mention below three kinds of element that exploit different physical phenomena.

Miniature mirrored louvres

This kind of system is made of fixed louvres that have curved slat profiles installed inside a double-glazing unit. By regular reflection, the curved slats perform a seasonally varying angular selection. In summer, sunlight is reflected back, but in the other seasons it is admitted and redirected. To work correctly, the profiles have to be adjusted according to site latitude and the slope of the area where the system is to be installed.

The main purpose of this kind of system is to control sunlight penetration. Miniature mirrored louvres are therefore not particularly efficient for increasing daylight factors. To date, such systems have been installed mainly to shade large glazed roofs.

Laser-cut panels

These elements are manufactured using a laser-cutting machine that produces a series of fine parallel cuts in acrylic panels. This process results in arrays of transparent parallelepipeds, with the cuts acting as a series of small reflectors with total internal reflection occurring at their surfaces (Figure 9.45). Usually these panels are also installed inside double-glazing units.

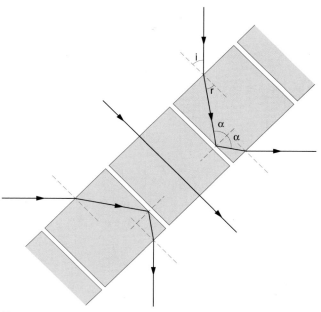

Figure 9.45 Reflection by total internal reflection in laser-cut panel.

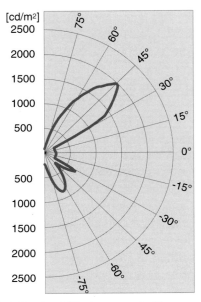

Figure 9.46 Indicatrix of diffusion computed for an acrylic laser-cut panel.

Figure 9.46 shows an indicatrix of diffusion computed for such a system positioned in a vertical opening. The redirection effect achieved by the cuts appears to be very effective. Thus sunlight or diffuse skylight can be redirected to the ceiling of the adjacent room. However, to reach remote parts of the ceiling, the redirected beam should be orientated at a lower altitude. This can be achieved either by using a panel with slightly tilted cuts or by installing the system in a tilted position (which ideally will vary with time).

An installation of laser-cut panels in a school classroom near Sydney, Australia, is shown in Figure 9.47. Several

designs, also for rooflights, have been devised with laser-cut panels and successfully tested (Figure 9.48).[17, 18]

In addition to their versatility and their potentially low manufacturing costs, laser-cut panels offer a nearly clear view out when viewed at near perpendicular directions.

Holographic optical elements

This type of element uses diffraction to achieve light redirection. Thin microscopic stripes forming a diffraction grating are printed on a transparent film. The gratings can be designed in order to perform a redirection for light incident at a specified angle. Incident light from all other directions remains unchanged. The films are finally laminated between two glass panes.

Since diffraction is dependent on the wavelength of the incident light, holographic optical elements can produce coloured fringes. However, modern design techniques allow these effects to be significantly reduced.

Several sunlight-redirecting or shading systems made of holographic optical elements have recently been built in Germany, where they were originally developed.[19]

Unlike the elements previously described, holographic optical elements are, to date, impossible to model using daylighting simulation programs. The design of systems that incorporate such elements thus requires physical modelling techniques.

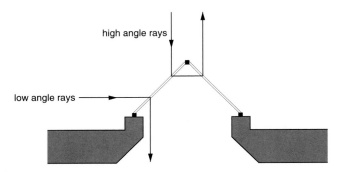

Figure 9.48 Laser-cut acrylic panel used in rooflight

Figure 9.47 Laser-cut panel installed in school classroom, Brisbane, Australia.

Light-guides

9.5.1 Sunlight-guiding systems

These systems aim to collect sunlight and channel it deep into the building. They comprise three components: an outside collector, usually located on the roof; a light-piping system that conveys the light beam into the building (Figure 9.3 shows various techniques used for this purpose); and, finally, a luminaire device that releases light into the interior space.

Since sunlight is a source of very high luminance, the light flux falling on a collector area of 1 m^2 around noon has the potential to provide an illuminance of 300 lux over approximately 65 m^2 of interior floor area (assuming a realistic efficiency of 20% for the whole system). Moreover, compared with an equivalent artificial lighting system, the higher luminous efficacy of sunlight means that this illuminance is obtained with significantly lower heat gains. These arguments were considered to be the main incentives for the development of the first sunlight-guiding systems at the end of the 1970s in the United States.[20] Since then, a limited number of sunlight-guiding systems have been installed.

However, a few systems are now available on the market – among them a Japanese system incorporating a sun-tracking collector comprising an array of fresnel lenses coupled to optical fibre cables, which can channel the light flux over long distances.

In Europe, sunlight-guiding systems have not yet attracted great attention. A few buildings have been equipped with them, mainly for aesthetic purposes in public areas.

Three main reasons explain the slow penetration rate of this type of technology:

- their relatively high cost

- their requirement for thorough and regular maintenance in order to preserve efficiency

- their poor performance in collecting and channelling diffuse light (due to the laws of physics).

This latter point restricts future development mainly to climates where cloudy conditions are very rare. In other climatic conditions, impressive and high-tech though they may appear, these systems are unlikely to play more than a decorative role.

9.5.2 Mirror light pipes

Recently lower-cost non-tracking devices referred to generically as mirror light pipes (Figure 9.49) have come on the market and enjoyed some popularity. Their performance in actual buildings is described in reference 21.

In one example, a room of approximately 3.6 x 2.3 m is illuminated by four 330 mm diameter vertical light pipes with clear external domes and diffusers in the lower apertures. The pipes were straight and 600 mm long. Measurements showed that in overcast conditions with an outside illuminance of 16 klux, the average internal illuminance was 177 lux, an effective daylight factor of 1.1%.

In another example, a room approximately 9 m square is illuminated by eight 330 mm diameter vertical light pipes with clear external domes and diffusers in the

Figure 9.49 Diagram of a sunlighting duct consisting of a pipe with a clear acrylic dome, polished aluminium interior and translucent ceiling diffuser.

transparent acrylic dome

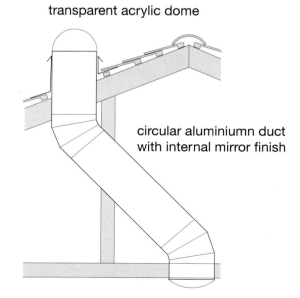

circular aluminiumn duct with internal mirror finish

diffusing light emitter

lower apertures but in this case a length between 8 and 12 m and containing up to four bends. In this case, with an outdoor illuminance of 80 klux, the average indoor illuminance was only 76 lux, an effective daylight factor of about 0.1%.

The results should come as no surprise since the upper aperture is not a concentrating device and therefore is no better than a normal rooflight. Indeed the first case, where the light pipes travel only through the thickness of the flat roof, would probably be as well served with a simple rooflight of the same total area, a square of side 600 mm. In the second case, a rooflight is not an option owing to the distance between the collecting aperture and the room. However, the performance is so low that it would be difficult to justify the expense of the installation.

The problem is that these systems are limited by the physics of their design. They cannot collect more light than is incident on the total area of the external aperture, which, since they are individually of small size, tends to be small in total. Secondly, this light has to be conducted by multiple reflections to the room, the number of reflections increasing with distance. For example, a ray at 45° to the pipe axis will be reflected at least L/D times, where D is the diameter and L is the length of the pipe. In the second case cited above, this means at least 24 times. An optimistic value for the reflectance of polished aluminimum or even aluminised plastic, with ageing, would be 0.9. The simple calculation

$$(0.9)^{24} = 0.079$$

indicates the nature of the problem.

Thus although these systems work in favourable conditions (short lengths and high external illuminance), where prevailing conditions are diffuse skies and longer pipe lengths are necessary, they are unlikely to be cost-effective and are not recomended.

9.5.3 Anidolic ceilings

Based on the anidolic zenithal collector presented in section 9.2.3, another experimental anidolic daylighting system has recently been developed with two aims: first, to facilitate its integration in buildings and, second, to limit glare for occupants located directly in the path of the light beam emerging from the exit aperture.

These requirements resulted in a design consisting of a collector protruding out of the facade that delivers the light flux via a guide installed as a false ceiling. An anidolic

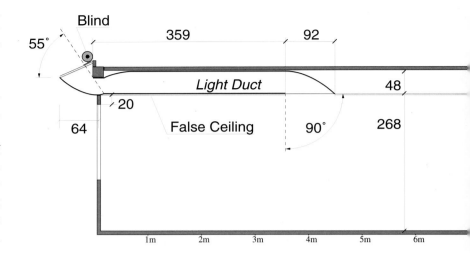

Figure 9.50 Cross-section of anidolic reflector and light duct integrated with ceiling in test room illustrated in Figure 9.51. (Giles Courret).

reflector then redirects the light flux into a horizontal exit aperture (Figure 9.50).

Since the collector is designed to admit diffuse light from nearly half the hemisphere, the laws of physics impose a minimum thickness for the light guide, i.e. about half the horizontal extent of the collector. A smaller thickness would inevitably result in a significant loss of the incoming light flux, since many rays would simply be reflected backwards. Conversely, a larger thickness would reduce the number of reflections each ray has to undergo and hence would improve light transmission. However, integration in a false ceiling would then become problematic.

The proposed design has been tested by simulation and subsequently built in a test module at EPFL, Switzerland (Figure 9.51). A retractable diffuse blind has also been installed in front of the entry aperture in order to limit excessive sunlight penetration.

Various measurements have been taken to compare the performance of the anidolic ceiling with a similar module equipped with double-glazed units only. In the back half of the module, the daylight factor is increased by a factor of 1.7. Using an automatically controlled artificial lighting installation (setpoint fixed at 300 lux), the anidolic ceiling required 33% less electricity than the reference module. A survey conducted with several occupants has indicated that this potential for energy saving may even be much higher when occupants manually control the artificial lighting. In the module

Figure 9.51 Outside facade of a test module equipped with an anidolic ceiling. Surrounding obstructions are not higher than 10° above the horizon.

equipped with the anidolic ceiling, the frequency with which they turned on the artificial lighting was seven times lower than in the reference module. This was as a result of the improved uniformity ratio.

An analysis of visual comfort conditions using the J-Index method (10.1.7) has also shown a significant improvement with the anidolic ceiling.[22]

Despite all these positive findings, an important question still needs to be addressed. Since the light-guide is made of large highly reflective aluminium sheets, the embodied energy of the system is quite high and this should be compared with the potential energy savings. Manufacturing aluminium sheets of this kind requires 200 kWh per square metre per mm thickness. This emphasises the need to investigate alternative reflective materials with less embodied energy before installing anidolic ceilings on a widespread scale.

Although they appear to be of major importance, such issues are to date rarely addressed when advanced daylighting systems are compared. They should in future be put on the daylighting research agenda and given high priority.

9.6 Summary

The intended purpose of Advanced Daylighting Systems is to improve the distribution and penetration of daylight into the room, in order to increase the usefulness of daylight. There are three broad categories of advanced daylighting systems: reflectors and light shelves, integrated elements in or close to the plane of the window, and reflective ducts with light-collecting and emitting elements at each end.

Systems which rely on direct sunlight show the best performance, reducing overall solar gain, whilst redirecting part of the direct flux into the back of the room, elevating the illuminance. These systems may also carry the advantage of smaller cross-sectional dimensions due to the high flux density of the direct beam. However they may perform poorly in diffuse conditions and are therefore only suitable for sunny climates.

Systems which re-direct significant amounts of diffuse light need to be more bulky due to the lower flux density, particularly in the case of ducted systems. However some systems, such as the classic light shelf, may reduce the front-to-back illuminance ratio, without actually increasing the illuminance at the back, thus improving the perceived daylighting of the room.

The specification of an advanced daylighting system should be preceded by a careful review of the required functions and the setting up of a conventional reference case. This will allow the improvement to be evaluated in order to justify (or not) the expense and possible inconvenience of an advanced system. In general, manufacturer's claims relating to performance should be regarded with caution.

Many systems have considerable visual impact from both inside and outside the building, and may present structural and maintenance problems. Furthermore, systems such as lightshelves and reflectors require the room to have specific optical and geometric properties. Thus the integration of systems with the architecture and structure must be given full consideration.

In evaluating the performance of an advanced system, the geometric property of re-directing and transmitting the incoming light (the indicatrix) should be used initially. Sophisticated CAD tools such as RADIANCE may also me used to simulate performance when the optical properties are known. Simple systems such as light shelves and louvers may be modelled physically, and some integrated elements such as holographic films, can be built into scale models.

References

1. N Ruck and S C J Smith, "The passive daylighting of building interiors", *Architectural Science Review*, vol. 31, 1988, pp. 87–98.

2. B Paule, J L Scartezzini, G Courret and L Michel, "DIANE daylighting project: dissemination of daylighting technology in the French part of Switzerland", *Proceedings of the European Conference on Energy Performance and Indoor Climate in Buildings*, Lyon, France, 1994.

3. G E Sweitzer, "Fixed-angle prismatic panel daylighting: daylighting distribution in perimeter office and Workshop areas", *Proceedings of the CIBSE National Lighting Conference*, 1992, pp. 275–282.

4. R Compagnon, J L Scartezzini J L and B Paule, "Application of nonimaging optics to the development of new daylighting systems", *Proceedings of the ISES Solar World Congress*, Budapest, Hungary, 1993.

5. R H Marshall and J L J Rosenfeld, "Design parameters for the optical and thermal performance of advanced glazing", *Proceedings of the UK-ISES Conference C69*, 1997.

6. R Compagnon, P Paule and J-L Scartezzini, "Design of new daylighting systems using Adeline software", *Proceedings Third European Conference on Solar Energy in Architecture and Urban Planning*, Florence, Italy.

7. R Mitanchey, G Periole and M Fontoynont, "Gonio-photometric measurements: numerical simulation for research and development applications", *Lighting Research and Technology*, vol. 27, no. 4. 1996, pp. 189–196.

8. G J Ward, "The Radiance lighting simulation and rendering system", *Computer Graphics Proceedings, Annual Conference Series*, ACM SIGGRAPH, 1994.

9. L Michel, C Roecker and J-L Scartezzini, "Performance of a new scanning sky simulator", *Lighting Research and Technology*, vol. 27, no. 4, 1995.

10. Ward, "The Radiance lighting simulation and rendering system".

11. J A Clarke, J W Hand, J Hensen, K Johnsen, K Wittchen, C Madsen and R Compagnon, "Integrated performance appraisal of Daylight Europe case study buildings", *Proceedings Fourth European Conference on Solar Energy in Architecture and Urban Planning*, Berlin, Germany, 1996.

12. Compagnon et al., "Application of nonimaging optics to the development of new daylighting systems".

13. G Courret, B Paule and J-R Scartezzini, "Anidolic zenithal openings: daylighting and shading", *Lighting Research and Technology*, vol. 28, no. 1. 1996, pp. 11–17.

14. CADDET, Japanese National Team, "Flat prism daylighting system", *CADDET Renewable Energy Newsletter*, 2/1997.

15. P J Littlefair, "Innovative daylighting: review of systems and evaluation methods", *Lighting Research and Technology*, vol. 22, no. 1, 1990, pp. 1–18.

16. P J Littlefair, *Designing with Innovative Daylighting*, BRE Report BR305 (Garston: CRC, 1996).

17. I R Edmonds, "Performance of laser cut light deflecting panels in daylighting applications", *Solar Energy Materials and Solar Cells*, vol. 29, 1993.

18. I R Edmonds, P A Jardine and G Rutledge, "Daylighting with angular selective skylights: predicted performance", *Lighting Research and Technology*, vol. 28, no. 3. 1996.

19. H F O Muller and M Kischkowest-Lopin, "First tests of daylight systems with holographic elements and light guide profiles", *Proceedings CISBAT 97*, EPFL Lausanne, 1997.

20. M Duguay, "Lighting with sunlight using sun tracking concentrators", *Applied Optics*, vol. 16, no. 1444, 1977.

21. L Shao, A A Elmualim and I Yohannes, "Mirror lightpipes: daylighting performance in real buildings", *Lighting Research and Technology*, vol. 30, no. 1, 1998, pp. 37–44.

22. G Courret, D Francioli and J-L Scartezzini, "Plafond anidolique: un nouveau système pour l'éclairage naturel latéral des bâtiments", *Proceedings CISBAT 97*, Lausanne, EFPL 1997.

10 Daylight, comfort and health

Interior of the
Hawkes' house in
Cambridge, UK.

Daylight, comfort and health

In the provision of a daylit environment three key aspects can be identified: visual performance, physiological conditions and visual quality.

Visual performance

Vision must be able to work as an efficient communication channel since we rely on it for more than 80% of the information we gather from our environment. Visual tasks need to be performed with accuracy, safety and at reasonable speed. These requirements imply various constraints on illuminance levels and luminance contrasts, mainly in the central part of the visual field on which attention is focused. The level of detail of the visual task and its complexity both affect these constraints. In addition, these constraints depend on the characteristics of the visual system of the person involved in the task. This introduces a large variability, as the visual system changes with age and varies considerably from one person to another; nearly 50% of the population need some kind of visual correction.

Physiological conditions

The visual field should not provoke excessive eyestrain or result in glare sensation. To meet this criterion, upper limits have to be set to luminance of directly visible light sources and luminance ratios in the whole field of view. Based on this criterion only, visual comfort cannot be considered as a perceptible sensation as it is not usually experienced by itself. In fact it results from the absence of stimuli recognised, consciously or not, as uncomfortable. Owing to its importance, many studies have dealt with this aspect, though mainly in the framework of artificial lighting. They concentrate on the identification and quantification of the phenomena that may produce uncomfortable stimuli. Their outcomes have been translated into practice by guidelines and even standards.

Visual quality

The field of view should present both aesthetic qualities and a certain degree of interest. These aspects play an important role in the judgement that people make of their lighting environment. Despite the fact that no physical methods exist to evaluate this aspect of the lighting design, some studies have tried to correlate measurable parameters with subjective impressions.

To meet all these requirements simultaneously presents a great challenge. At the design stage, important factors often remain unknown (e.g. type of activity, furnishing locations, colours). The dynamic lighting conditions encountered in daylit space adds even more difficulty. These lead inevitably to assumptions and compromises. To base these decisions on sound arguments is of prime importance to ensure successful daylighting designs, particularly because clients currently consider visual comfort criteria with greater priority than energy savings.

It is commonly recognised that access to daylight is very much appreciated by occupants, although claims that daylit workplaces have a marked positive impact on productivity have not been scientifically validated so far. However, there are physiological effects of daylight that are strongly associated with health and well-being. Furthermore, there are well-known disbenefits associated with the artificially lit environment. These health issues are dealt with in more detail in 10.2.

Visual comfort

10.1.1 Light perception

On the back surface of the eye, the transformation of light into neural activity is carried out by light-sensitive receptor cells (photoreceptors). The retina contains two types of photoreceptor: the rods (approximately 120 million) and cones (approximately 6 million). Rods are able to detect light at low luminance levels (<1 cd/m^2) and provide low-acuity monochrome vision called 'scotopic vision'. There are three types of cones, which are sensitive to either blue, red or green light. The cones are principally located in a very limited area of the retina called the fovea. This is the part of the retina where the eye keeps the viewed image in focus by its accommodation mechanism involving adjustments of the lens curvature through appropriate contraction of the ciliary muscles (Figure 10.1). Owing to their location and light sensitivity characteristics, the cones provide high-acuity vision called 'photopic vision'. For luminance around 1 cd/m^2, the two kinds of vision compete simultaneously, and this domain is called 'mesopic vision'.

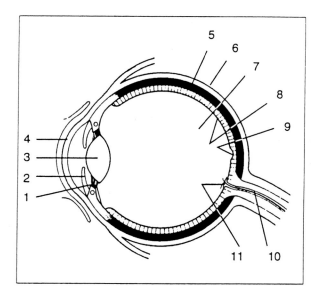

Figure 10.1 Cross-sectional diagram of the human eye: (1) ciliary ligament, (2) iris, (3) lens, (4) cornea, (5) choroid, (6) sclera, (7) vitreous body, (8) retina, (9) macula lutea, (10) optic nerve, (11) blind spot.[1]

As shown in Table 10.1 the perceptible luminance range of the human visual system spans a very large interval. Excluding view of the sun either directly or after specular reflections, the visual experience in daylit spaces

Table 10.1: Range of perceptible luminances. The luminance range usually experienced in daylit spaces is from 1 to 10^4 cd/m^2.

	Luminance (cd/m^2)	
Electronic flash	10^{10}	
The sun surface by clear sky	10^9	Damaging
	10^8	
Filament of a halogen lamp	10^7	
Filament of an incandescent light bulb	10^6	
	10^5	
White paper in sunlight	10^4	Photopic vision
White paper under an overcast sky	10^3	
White pixels of VDU screen	10^2	
Comfortable reading	10	
	1	Mesopic vision
	10^{-1}	
White paper in moonlight	10^{-2}	
	10^{-3}	Scotopic vision
White paper in starlight	10^{-4}	
	10^{-5}	
Weakest visible light	10^{-6}	

appears rather limited as the luminances that may be encountered span only over five orders of magnitude across the mesopic and photopic domains.

After being detected by the rods and cones, light is transformed into a pattern of neural activity that travels towards the brain through the optic nerve. As the latter contains only around 1 million fibres, the retina has to condense and reorganise the information gathered from the 126 million photoreceptors. This first information-processing task occurs in the eye, in the retinal ganglion cells. The resulting visual map sent to the brain highlights, for instance, areas where there are changes of luminances, such as edges of objects. On the other hand, contiguous areas of the visual field that have similar colours and slowly varying luminances are merged with less detail. This strategy clearly preserves the features of the image that are relevant for a correct interpretation of the perceived objects and space. In recent years, neurobiologists have shown that, in spite of our unified visual experience, specific image attributes are then analysed in different areas of the brain.[2]

10.1.2 Visual adaptation

To cope with the fact that the optic nerve fibres can only transmit signals of limited range (perhaps 1:100) compared with the detectable luminance range (Table 10.1), the eye has to adapt its sensitivity to the actual perceived visual field. The pupil's diameter is first adjusted by the iris to control the amount of light entering the eye. A second mechanism, named adaptation, plays an even more important role: the retinal ganglion cells themselves adapt their response levels to the average illumination of the retina. The resulting effect is that, at any one time, the perceptible range of luminance spans over three to four orders of magnitude between two thresholds: a lower boundary below which no luminous sensation will be experienced, and an upper boundary above which glare sensation will occur (Figure 10.2).

This process is very often experienced. For instance, when looking into a building from outdoors, the rooms seen through the windows usually appear barely visible. This happens because the outdoor average luminance level exceeds 1000 cd/m^2 and the visual adaptation accordingly adjusts the shadow limit between 10 and 100 cd/m^2 which are luminance levels typically encountered in a daylit room. When inside the same room, the

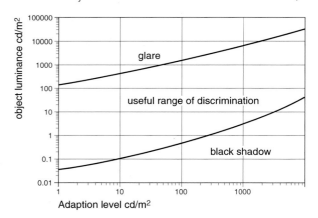

Figure 10.2 The resulting effect of visual adaptation: depending on the adaptation level, the eye adapts its sensitivity so that a limited range of luminance is perceived. Luminances lying below the lower bound are perceived as black shadows and, on the other hand, luminances lying above the upper level appear as glaring. The limits curves shown are in no sense sharp boudaries; loss of shadow detail and glare are gradually experienced around these limits.

visual system will then adapt to this level so that this range of luminance will finally fall in the useful range of discrimination where luminous sensations are perceptible (Figure 10.2).[3]

The same phenomenon explains why a patch of sky that appears acceptably bright when seen from outdoors may cause glare when seen from inside through a window.

As shown from these examples, the visual adaptation mechanism has a profound implication: the luminance (measured in cd/m^2) does not relate uniquely to the apparent brightness. This means that, depending on the average luminance level of the whole field of view (expressed as an adaptation level), the same luminance can be perceived from dull to glaring. For example, the moon appears much brighter during nighttime (low adaptation level) than during daytime (relatively high adaptation level) while its luminance remains the same in both cases.

A decrease of the adaptation level, for instance when moving from a bright location to a darker one, is not immediate but takes some time: after 0.1 seconds the eye is able to decrease its adaptation level by a factor of 3, but a further decrease needs a few minutes more. Full adaptation to a level falling in the scotopic range may take from 30 to 60 minutes. This can be experienced for instance when entering from daylight into a darkened cinema where the film has already started: apart from the screen we can see very little at first, but after a while the surroundings gradually become visible. Conversely, a change of adaptation from dark to bright happens more rapidly and is fully achieved after a few minutes. These transient effects impose a limiting criterion for the design of adjacent spaces between which frequent movement is expected; their average daylight factor should not differ by more than a factor of 3. This is also in accordance with the Lynes criterion (4.3.3) as it prevents too large adaptation level changes in a single daylit room where the gaze may move from one place to another.

10.1.3 Apparent brightness

The concept of apparent brightness originated in the 1930s. Since then several researchers have used various experimental procedures to establish relationships between luminance, adaptation level and subjective sensation of apparent brightness. One such relation is depicted in Figureure 10.3 for various adaptation levels; luminances are mapped onto an arbitrary scale of apparent brightness magnitude. This scale accords well with subjective sensation, in that a magnitude 2X appears 'twice as bright' as a

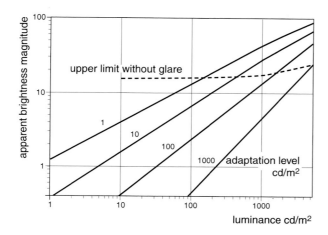

Figure 10.3 Apparent brightness vs. luminance and adaptation level as proposed by Hopkinson. The upper limit without glare has been superimposed as a dashed line. A log–log graph is used for its property to transform power law curves into straight lines.[4]

magnitude X. Furthermore, if asked to numerically quantify their subjective sensation of brightness, people on average choose numbers of this order.

The corresponding mathematical expression is a power law of the form

$$(\text{Apparent brightness}) = k(A).L^{b(A)}$$

where: $k(A)$ and $b(A)$ are functions of the adaptation level A and L is the absolute luminance.

This is not surprising, as many experiments have demonstrated that many other physical stimuli (e.g. sound loudness, temperature, vibration) also relate to their corresponding human sensation estimation through power laws. The exact relationship is still subject to debates, mainly because experimental procedures widely differ among researchers.

This relationship in fact relates the brightness we see to the photometric quantity that can be measured (luminance). Important considerations can now be deduced. Suppose a designer wants to highlight a part of a wall by doubling its apparent brightness. The adaptation level is typical of a daylit space and lies around 100 cd/m². Figure 10.3 shows that to double the apparent brightness the corresponding luminances need to be increased by a factor of 2.5. For lower adaptation levels, as encountered in offices with VDU workplaces, this factor even increases up to 3. How to achieve this goal? A first solution would be to cover this part of the wall using a paint with a reflectance 2.5 or 3 times higher than the original wall colour. If the reflectance of the latter is already higher

than 30% then it is physically impossible to find a paint that will be 2.5 or 3 times more reflective. Another solution would be to increase the overall illumination level by the same factor, either by changing all reflectances and/or by increasing the window area. This solution is not really equivalent to the former since the luminance of the whole space will increase, and not only the part of the wall under consideration. In fact, this strategy is even less likely to succeed since it simultaneously increases, the adaptation level. The resulting effect is depicted in Figure 10.4: depending on the initial adaptation level, the relative change of apparent brightness is given for a range of luminance multiplying factors.

Figure 10.4 shows that the doubling of the apparent brightness is not easy to achieve. Even if multiplied by a factor of 5, the luminances in the typical daylit space of 100 cd/m² will be perceived 1.5 times brighter only. The visual system shows an obvious trend to maintain the apparent brightness nearly constant even if the illuminance levels vary on a large scale. This phenomenon plays a key role in our ability to recognise spaces and objects around us in spite of our continuously varying luminous environment. It has also an important consequence in that even a skilled lighting engineer cannot simply rely on his visual system to estimate luminance or illuminance levels in a room; relative errors of 50% are quite common.

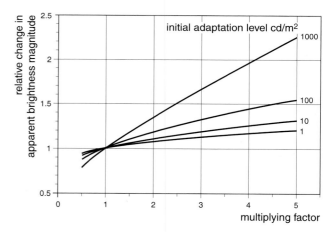

Figure 10.4 Relative change in apparent brightness when scaling luminances by a constant factor for the whole field of view. The curves are labelled by the initial adaptation level (i.e. before the luminance is scaled). The final adaptation level is simply the product of the initial level and the multiplying factor.

10.1.4 Lightness constancy

The interpretation of the visual field occurring inside the brain also affects visual perception. By identifying the locations of the light sources, the brain is able to roughly assess the prevailing illuminance distribution pattern (e.g. the illuminance falling on a surface increases when approaching a window; surfaces that do not face towards windows are less illuminated). Some kind of assumptions taken from common sense or gained from past experiences are extensively used at this stage (e.g. a surface is usually painted in a uniform colour over its entire area; daylight comes downward from the sky). Combining these clues, the brain is able to distinguish between the reflectance of an object and the illumination falling upon it.

For example, a uniformly white ceiling lit from a side window will look white all over its area in spite of a variation of more than 10:1 in luminance. A grey wall opposite the window receiving light almost normal to its surface may have a higher luminance than the adjacent patch of ceiling receiving light only at a glancing incidence. Nevertheless the grey wall will still appear grey compared with the white ceiling even though their luminances are in reverse order. This ability resulting from the visual interpretation process is called 'lightness constancy'. It is further enhanced by 'colour constancy' (6.2.4). These mechanisms are both of prime importance to enable a reliable perception of objects and their surrounding space.

10.1.5 The visual field

The visual field is bounded by a cone of approximately 140° apex. It comprises three distinct parts that have quite different characteristics (Figure 10.5). In the centre a restricted zone bounded by a cone of 1° apex is called the 'area of central vision'. Light rays perceived from this range of directions reach the retina on the fovea, where the cones are located at their maximum density. Therefore this area provides the most acute vision, and it is instinctive to shift the gaze until the visual task falls exactly in this central area. To allow the brain to build up a sharp picture of larger portions of the visual environment, the eyes need to move rapidly around this direction. These movements are performed very precisely and rapidly by the eye muscles. (A shift of 10° may be accomplished in about 0.04 seconds.) A second circular zone delimited by a cone of 60° apex encircles the central area. Here the light is perceived mainly by rods that make vision become progressively blurred and indistinct as the incidence angle from the centre increases. This zone is called the 'ergorama' since it usually embraces a portion of the space containing various objects that are necessary to perform a working task (e.g. a computer workstation with its screen and keyboard close to paper documents). Finally the 'panorama' fills the outer part of the visual field. Its extent is limited by the nose, forehead and cheeks. In the panorama

Figure 10.5 RADIANCE pictures of the EOS building, Lausanne, Switzerland. Fish-eye views of the visual field as perceived from a workplace. On the left: a reading task with a line of sight 45° below the horizontal plane. On the right: a VDU workplace with a line of sight 12.5° below the horizontal plane. The area of central vision is located at the centre of the black circular target. The surrounding ergorama is shaded with a yellow colour. The parts of the visual field covered either by the nose, forehead or cheeks appear in blue colour. Lateral sides of the panorama shaded with a blue colour are perceived by a single eye only.

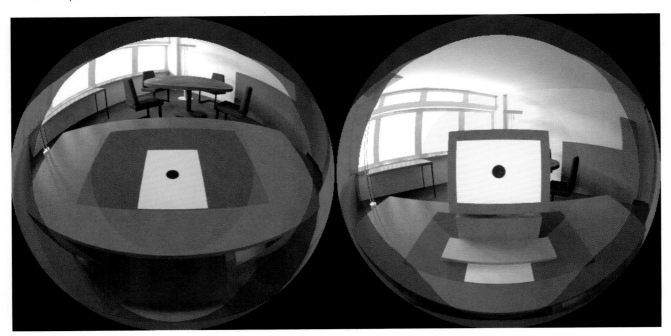

objects are hardly noticeable unless they move. This divison of the field of view is used mainly for luminance ratio criteria that will be presented later in this chapter.

As shown in Figure 10.5 the orientation of the task (either horizontal or nearly vertical) has an important impact on the type of surfaces seen in the different parts of the field of view. For a horizontal reading task, both the ergorama and the panorama embrace mainly horizontal surfaces such as the workplane and the floor. The forehead prevents the direct vision of the ceiling and the upper part of the window (in this case occupied by a lightshelf). For a VDU task, a large portion of the field of view is made of vertical areas such as the VDU screen, walls and windows. In addition the ceiling and the upper part of the windows are located inside the panorama.

These examples illustrate the duality of the problem when dealing with visual comfort issues: in the first case the furnishing occupies the most part of the visual field, but this is not really under the architect's control at the design stage. Conversely in the second case the surrounding space covers a much larger fraction of the visual field. Therefore architectural design of the whole room becomes of prime importance.

In the recent past, office work involved mainly horizontal paper-based tasks. This explains why numerous lighting standards concentrate on workplane illuminance levels criteria. Nowadays approximately 50% of European office workers regularly use VDU workstations, and this proportion will certainly continue to rise. This profound change means that, in addition to illuminance levels, luminance distributions in the field of view have to be carefully considered.

10.1.6 Visual performance

Apart from the quality of the visual system (eyesight) that largely varies between individuals, and other psychological factors like motivation or concentration, the accuracy and speed (i.e. the visual performance) with which a visual task is performed is related to the following factors:

- the apparent size of the details to be perceived on the task and the luminance contrast between details and background

- the illuminance of the task

- the visual fatigue state

The area of central vision is of concern for the first factor, the ergorama for the second, while the whole field of view affects the last. All of these factors can be controlled to some degree through adequate design, as described below.

Apparent size

It is of course common experience that the larger a detail is, the easier it is recognised. When working on a VDU workstation, the character size can easily be altered. Conversely when the size of an object is not under control (e.g. a book), its apparent size can still be changed by approaching the eye closer to the task and so increasing the area of the image projected on the retina. This may need constrained postures (e.g. students bending their necks awkwardly while reading) that could become physically painful if the workplace was not ergonomically adapted.

When a single task has to be seen by many observers simultaneously (typical in educational contexts), the purpose of minimising the distance separating the observers from the task can largely affect the shape of the space. Figure 10.6 illustrates such an extreme case from an era long before any technical options were available. Video recording near the task, directly projected either on a large screen or multiple VDU displays, constitutes the contemporary technical response to the same problem. However, it does not always constitute a convenient solution since video display systems provide rather low-luminance pictures, and it therefore requires a low general illuminance level that could make simultaneous writing tasks more difficult if not impossible.

An increase of the luminance contrast between the details to be perceived (e.g. characters) and the background (e.g. paper or VDU) greatly improves the acuity the observer can reach. Expressed as luminance ratios between details and background, contrasts of 6:1 (bright details on a dark background, so-called 'positive contrast') or 1:6 (dark details on a bright background, so-called 'negative contrast') are already sufficient for good recognition, while ratios below 3:1 or above 1:3 will seriously deteriorate the acuity. These contrasts originate from the task itself, either from reflectance differences between details and background (e.g. a dark ink on a white paper) or from luminance variations between the miniature pixels forming a screen display. Except for the latter, these are not commonly adjustable parameters.

Task illuminance

Task illuminance plays an important role as it is of course required to reveal the contrasts on visual tasks that are not

Figure 10.6 Anatomical theatre, Uppsala, Sweden (1662). The circular arrangement of the seats on a very steep slope ensures that the distance separating the observers from the task is kept at a minimum so that details on the task appear large enough to be clearly seen.

self-luminous. In addition, when the luminance of the task background increases, the acuity also increases since light rays focus more precisely on the fovea when the pupil becomes smaller. Acuity increases with task illuminance, reaching a maximum at around 1000 lux. However, for people with normal vision the rate of increase in acuity falls off above 100 lux. This explains why lighting standards prescribe task illuminances usually above 100 lux, up to higher values according to the fineness of detail.

Task illuminance sometimes leads to unwanted effects when bright parts of the incoming light flux are specularly reflected directly towards the eyes. This may occur whenever the task-supporting media or the details themselves exhibit specular reflectance components (e.g. glossy paper, VDU, glossy ink). The reflections of the light sources on these surfaces (so called 'veiling reflections') superimpose high luminances on the lower task luminance. This is first recognised as a source of distrac-

tion. In addition the perceived contrast between details and background may be largely reduced. In worst cases, the visual task can no longer be achieved. Owing to their glass surface VDUs are prone to veiling reflections, especially when installed in highly daylit spaces (Figure 10.7).

Figure 10.7 Waterloo International Terminal, London, UK. Veiling reflections of the glazed roof on a VDU seriously impair the visibility of information it is intended to display. The dark screens positioned around the four edges of the VDU are not large enough to prevent any veiling reflection. A slight inclination of the VDU towards the floor would be more efficient.

To avoid this impairing phenomenon, called 'glare by reflection', no light source should be able to reflect on the task directly towards the eyes. This imposes restrictions on the relative positions of the tasks and the light sources. Figure 10.8 depicts how an 'offending zone' can easily be defined. Ideally, neither window nor artificial light source should be located inside the offending zone

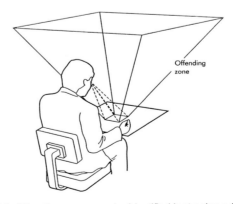

Figure 10.8 Offending zones can be identified by tracing a few rays in the reverse direction that they would naturally follow (this is known as a 'backward ray tracing method'). Starting from the eye the rays are first sent towards the task. By assuming the task is purely specular the corresponding reflected rays are then traced up to the first surface they reach.[5]

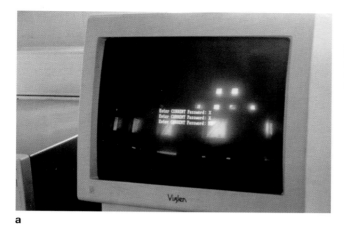

a b

Figure 10.9 Queen's Building, Leicester, UK. Veiling reflections on a screen that contains windows in its offending zone. The worker sitting in front of the VDU does not completely hide the view of the windows.

of each visual task (Figures 10.9 and 10.10). In addition, if sunlight is tolerated into the space sunny patches

should not reach any offending zone in the presence of observers. These requirements imply a detailed knowledge of the visual tasks layout at the design stage. Since this is not always possible, an adequate daylighting design should at least respond to highly probable task locations and identify locations unsuitable for visual tasks. These issues have to be closely discussed between designers and clients throughout the design process.

A loss of contrast sensitivity can also arise inside the eye itself when the visual field comprises areas of high luminance compared with the prevailing adaptation level. This is due to the light scattered by opaque particles that accumulate with age in the vitreous humour (Figure 10.1), and to the adaptation mechanism that diminishes the apparent brightness of the object of regard. The resulting veiling effect decreases rapidly with increasing angular distance between the high-luminance source and the direction of viewing. Therefore surfaces that may exhibit high-luminance spots should be avoided in the ergorama. For instance glossy painted desks may cause veiling problems when windows are located within their offending zone. Windows may still be tolerated within the ergorama as long as the view towards these directions only reaches external obstructions of relatively dark colours (e.g. vegetation).

Glare by reflection and the loss of contrast sensitivity due to veiling effects occurring in the eyes lead to what is called 'disability glare' since under these circumstances visual tasks become difficult if not impossible to perform.

There are two further reasons to balance luminance in the ergorama. First, acuity varies with the luminance of the task's immediate surroundings (the ergorama). The optimum luminance that gives greatest visual acu-

Figure 10.10 Ludwig Museum, Köln, Germany. Owing to their high location, windows do not fall in the offending zone of each painting.

ity is less than or equal to that of the task itself, but greater than one third of that value. The decrease in visual acuity is more pronounced if the luminance of the ergorama is raised above that of the task than if it falls below the lower boundary. A slightly lower luminance in the ergorama also helps to maintain attention focused on the task. The second reason relates to the transient adaptations that occur when moving the gaze towards various parts of the ergorama: if luminance ratios are too high the visual performances decrease since brief delays are experienced while the eyes continually readapt.

Visual fatigue

Visual fatigue results if the above requirements concerning either the task or its illuminance are partially or fully unsatisfied for a prolonged period. The symptom is an increase of the distance of the near point. Since visual tasks are generally located between 0.3 and 0.7 m from the eyes, the temporarily receded near-point may well fall in this range. Hence the visual performance deteriorates.

To enable the ciliary muscles to have periods of rest while performing a visual task, the view field should contain some kind of 'escape regions' that provide unobstructed views of non-glaring surfaces located at least 3 m away. This is certainly something we instinctively search for when choosing a workplace or adjusting its position or layout. Furthermore it appears that this requirement predominates over the luminance balance criteria presented above. For instance it is quite common to observe VDUs installed in front of windows or internal spaces (Figures 10.11 and 10.12) where daylight frequently is excessive. This shows the need for

Figure 10.12 Irish Energy Centre, Dublin, Ireland. Although it is essential to prevent the direct view of the large rear window, the partition over the desk is so close to the workplace that it does not offer any "escape region" in the field of view. Hence the VDU has been moved close to the window.

the designer to devise means of luminance controls such as lightshelves or movable solar protection.

10.1.7 The J index

A new method has recently been developed to quantify visual comfort in terms of visual performance in the workplace. The basic principle is to measure the difference between the maximum acuity, A_{max}, a person can reach in ideal lighting conditions and the acuity, A, obtained in the workplace. For this purpose a discomfort index J is computed as:

$$J = \frac{A_{max} - A}{A_{max}}$$

J varies according to the lighting conditions in the field of view between 0 (ideal situation) and 1 (worst lighting conditions due to either disability glare or excessively low luminance levels).

A_{max} characterizes the visual system of the person for which the analysis is performed. This value is obtained from a simple visual test. The acuity A is deduced using three measurable parameters that have physiological effects on the visual system: the luminance contrast between the target (typically the text) and the background of the visual task (a sheet of paper or a VDU screen), the luminance contrast between this background and its surrounding (i.e. the remaining part of the visual field), and the illuminance at eye level. These parameters are obtained either from measurements performed in the workplace or from computer-generated pictures (see Figure 10.5).

For a specific task, a required acuity A_e can be specified and gives a corresponding threshold value J_e above

Figure 10.11 APU Learning Resource Centre, UK. A VDU workplace with distant surfaces located in the field of view enables resting accommodation for the eyes. Visual performance may be reduced owing to the high-luminances in the ergorama compared with that of the task.

which discomfort will be experienced. Note that with this approach discomfort is assumed to be directly linked to the degradation of visual performances. Thus this method is devoted mainly to the ergonomic analysis of workplaces.[6]

From the measured parameters a set of J index values are obtained and compared with the threshold value J_e. This enables personalised recommendations to be made regarding the illumination of the workplace. For a daylit workplace, its location and/or orientation relative to the windows are often modified accordingly.

To enable more general recommendation, this method is also able to take into account the variability of the visual system performances over the whole population. This allows the calculation of the predicted percentage of dissatisfaction (PPD), which can even be split into four categories:

- PPD<: dissatisfaction due to a lack of light

- PPD>: dissatisfaction due to an excess of light

- PPD~: dissatisfaction due to inappropriate contrasts

- PPDT: the total percentage of dissatisfaction

For instance, the J index procedure applied for the two workplaces illustrated in Figure 9.5 gave the following results:

	Horizontal reading task	VDU
PPD<	11%	4%
PPD>	0%	0%
PPD~	2%	2%
PPDT	39%	31%

Note that the total percentage of dissatisfaction is always higher than 25%. This results from the fact that this fraction of the population cannot perform reading tasks with a sufficiently high margin between the best acuity they can reach and the required acuity. However, this does not mean they are totally unable to perform such a task. For the most severe cases this may have postural implications or imply means of optical enlargement of the task details.

10.1.8 Discomfort glare

The presence of high luminance contrasts in the field of view may cause discomfort sensation even if little or no decrease in visual performance is observed. This phe-

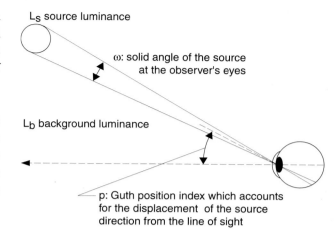

Figure 10.13 The parameters taken into account to compute discomfort glare indices.

nomenon is known as 'discomfort glare'. It originates from an instability in the control mechanism of the visual system. On one hand the highly stimulated parts of the retina receiving illumination from the glare source demand to close the pupil while, on the other hand, the less stimulated parts of the retina demand the reverse.

Many studies conducted more than 50 years ago have shown that the magnitude of the discomfort glare sensation is directly related to the luminance of the glare source and its apparent size as seen by the observer. In addition, discomfort is reduced if the source is seen in surroundings of high luminance. This is due to the adaptation phenomenon (10.1.2), and is frequently experienced by car drivers: headlights often appear glaring at night but remain barely noticeable during the day.

The glare sensation is also reduced the further the glare source is off the line of sight (Figure 10.13). This is again something often experienced on sunny days: the glare sensation resulting from the direct view of the sun rapidly decreases when the gaze moves away from its direction.

All these findings have been translated into empirical formulae in order to quantify the discomfort glare sensation using a so-called 'glare index' scale. Various slightly different versions have been proposed. The latest one recently defined by the CIE is called the 'unified glare rating' (UGR). Since more than one glare source may be present, the formula involves a summation:

$$UGR = 8\log\left(\frac{0.25}{L_b}\sum\frac{L_i^2\omega_i}{P_i^2}\right)$$

where:

- L_i is the luminance of the glare source i

- w_i is the solid angle subtended by the source i

- P_i is the position factor

- L_b is the mean luminance of the remaining parts of the field of view

UGR values below 10 mean imperceptible glare. Glare becomes really uncomfortable for UGR values above 22. Values above 28 denote intolerable glare sensations. A variation of a single unit on this scale is just noticeable. A three-unit variation represents a difference that is always recognised. Therefore a decrease of three units on the glare index scale is considered as a worthwhile improvement when trying to limit discomfort glare.

The UGR system is used mainly to make comparisons between various artificial lighting installations. Field studies have been conducted to estimate adequate limits to glare index values for various building types. In addition, the occupation times, the degree of freedom that occupants have to turn their gaze away from glaring light sources, the nature of the visual task to be performed and the degree of attention it requires were also taken into account. Three broad categories have been defined:

- environments where no glare at all is permissible; glare index limit 10

- environments where glare must be kept to a minimum; glare index limit 13

- environments where different degrees of glare are tolerable; glare index limits between 16 and 28

Specific glare index limit values are defined by various national standards. For daylit spaces, the evaluation of glare indices is less straightforward. The main difficulty lies in the precise distinction between the glare sources and the background in the perceived field of view. The sky seen through a window and other bright surfaces located either outside or inside the room may be considered as glare sources. The dynamic nature of natural illumination makes this distinction vary over time: the same clear area may become glaring when directly lit by sun rays but otherwise remain part of the 'background'.

Compared with artificial lighting installations, where each luminaire occupies a small portion of the field of view, the glare sources identified in a daylit room are often much larger. To account for these differences, a specific 'daylight glare index' (DGI) has been defined. Its mathematical definition (known as the Cornell formula)

is slightly more complicated than that for the UGR but remains of very similar nature.

Field studies have shown that occupants have a greater tolerance of mild degrees of glare from the sky seen through windows than for a comparable artificial lighting system with the same value of glare index. This means that the correlation between sensation and glare index is modified slightly to handle DGI values. Up to 16, glare remains imperceptible. DGI values above 24 denote an uncomfortable glare sensation, becoming intolerable above 28.

It has also been demonstrated[7] that the DGI value computed for a daylit room is nearly independent of both the window size and the location of the occupant as long as the window area is greater than 2% of the floor area. This means that, for given sky conditions, a daylit space has a relatively fixed glare character. Thus it is more interesting to find out how design parameters can decrease the DGI value than to calculate its exact value.

The major controlling factor is the glare source luminance. To obtain a significant decrease of the DGI value (−3 units), the luminance has to be divided by a factor of 4. Shading systems are usually used to achieve this goal. In buildings with high thermal inertia (typical of passive solar buildings) it has been demonstrated that movable shading systems are more frequently used to control discomfort glare than to avoid overheating.

This requirement for limited luminance directed towards the occupants from the window openings is one of the major justifications for the development of innovative daylighting systems.

To compare their potential for limiting discomfort glare, their indicatrix of diffusion should be examined (9.2.3). Two systems can be considered as significantly different in terms of direct discomfort glare if their luminances emitted towards the zone defined by altitude between 0° and −50° differ by a factor of 4 or more.

Glare sources located in directions making angles of greater than 50° with the line of sight (in the horizontal plane) have little influence on the DGI. This fact constitutes another effective method to limit discomfort glare. The orientation of the workplace can be adjusted in order to maintain the most probable line of sight off the directions of the openings by at least 50°. Such measures can even be more effective than the use of shading devices (Figure 10.14).

Figure 10.14 Tractebel Building, Brussels, Belgium. The back-illuminated translucent blind is a glare source lying well within the 50° limit of line of sight to the VDU screen.

When the occupants have no means to screen off the glare sources, or no freedom to displace their principal line of sight, they often use alternative practical methods. Large plants located close to the facade, cardboard screens or posters hung on windows are typical strategies (Figure 10.15). When frequently observed in a daylit building, these occurrences should be considered as strong indicators that discomfort glare is above acceptable limits, even if no complaints have been registered.

Since discomfort glare results from excessive contrasts, this sensation can also be controlled by increasing the luminance of the whole field of view. For instance, changing the room mean reflectance from 0.4 to 0.6 causes a decrease of −2 units on the glare index scale.

It has also been shown that a glaring source that is not sharply delineated against the background causes less glare than a sharp edge. This finding suggests another way of mitigating discomfort glare, by 'contrast grading'. All potential glare sources should be directly surrounded by areas of relatively high luminances. This can be achieved by using light matt colours for window frames and glazing bars. Splayed window sills, heads and lateral sides also reduce sharp edge-contrast effects. Care should be taken, however, to ensure that these surfaces do not themselves become secondary glare sources. This risk is particularly high with specular finishes.

Windows installed in more than one wall can also help to reduce contrasts. In addition to increasing the

global luminance, and hence the surround adaptation luminance, they can mutually increase the luminance of their respective walls, which contributes to the contrast reduction (4.3.6).

The daylight glare index does not fully take contrast grading into account. Consequently this way of reducing discomfort glare does not help to meet statutory DGI limit values. However, it is well recognised that contrast grading is an effective method for limiting discomfort glare sensation. Furthermore it is directly under the architect's control.

10.1.9 Occupants' preferences

Visual comfort does not only rely on the satisfaction of physiological lighting requirements. Subjective reactions of the occupants are of at least equal importance. This section summarises the outcome of a series of German surveys conducted on a large sample of dwellings.[8] Correlations have been established between measurable

Figure 10.15 The Scottish Office, Victoria Quay, Edinburgh, UK. A poster hung on a window to counteract glare.

parameters and occupants' preferences. Two issues are of particular interest: the provision of daylight and sunlight, and the size of the windows.

Regarding the provision of daylight, an empirical relation between the mean daylight factor and the occupants' subjective rating has been found. The rating is recorded on a satisfaction scale ranging from −2 (very bad daylit environment) to +2 (very good daylit environment). It appears that the satisfaction falls off rapidly for mean daylight factors below 0.9%. Consequently this level has been adopted as the mandatory minimum in the DIN 5034 standard *Daylighting in indoor rooms*.

Sunlight penetration is also very much appreciated. In order to satisfy at least 70% of the occupants, it has been established that, at the equinox, the centres of the windows should receive direct sunlight for four hours or more. Sunpath projections (2.1) are convenient tools to ensure that this requirement is met. This rule has been deliberately made simple, and its use should be restricted to latitudes and climates similar to Germany, where the underlying surveys have been conducted. This issue is usually addressed by relevant national standards.

The surveys have demonstrated that satisfaction increases steadily with higher window areas. This means that occupants do not complain about too large windows. A lower limit has been defined in order to satisfy 70% of the occupants, where the transparent part of the window should amount to 30% of the window wall area or 16% of the floor area.

The visual connection to the outside is a major factor affecting the preferred size of the windows. Preference criteria regarding width and height of windows have been identified. The width of the transparent part of the window should always exceed 55% of the wall where it is installed. The satisfaction gradually falls off if the same glass width is distributed among many smaller windows, and if their distance apart increases.

Regarding the height of the transparent part of the windows, an absolute lower limit fixed at 1.3 m appears to prevail.

Apart from window sizes, their positions also have great impact on occupants' satisfaction. A clear view of the skyline should be provided as often as possible. This is not surprising since the strongest jumps in luminance and usually the sharpest changes in geometry occur at the skyline. Thus the visual information is concentrated at the skyline, which, inevitably, becomes very attractive to the eye.[9]

A clear view to the outside is problematic when daylighting systems are installed. Usually a conventional window is located at the eyes' height and the daylighting system remains located in the upper part of the opening. The essential requirement for a clear view also stimulates the development of daylighting elements that keep their transparency for a large range of directions.

Although the findings presented above result from surveys conducted in dwellings equipped with conventional windows, they are likely to remain valid and applicable for other building types and also for windows equipped with daylighting systems. Post-occupancy procedures (Chapter 13) allow more detailed analysis of occupants' satisfaction regarding innovative daylighting designs. However, the present lack of a sufficient number of well-documented post-occupancy studies makes the generalisation of their outcomes difficult.

Recent work carried out in Cambridge has confirmed the influence of non-technical factors on perceived visual comfort, and has cast some doubts on the validity of existing daylight glare indices when applied to real (non-laboratory) conditions. This work is described in detail in 13.4.

10.2 Daylight and health

In common with most other higher organisms, humans depend on exposure to daylight to activate a wide range of physiological functions. In essence there are two aspects:

- the intensity of daylight exposure

- exposure specifically to the ultraviolet (UV) component of daylight.

10.2.1 Intensity of daylight exposure

Circadian rhythms

Doses of strong light are needed each morning to prompt the pineal gland to switch off production of melatonin. Receiving information on light levels from the retina, the pineal gland is a pea-sized organ at the base of the brain that performs several important regulatory functions, chiefly through the release of melatonin into the bloodstream during the hours of darkness. In this way, a wide range of body organs, including the brain, exhibit diurnal (or circadian) cycles of activity.[10, 11] Thus the release of melatonin at night makes us feel sleepy, suppresses our endocrinal system (to reduce stress), and damps down other functions that might interfere with sleep. All of these effects are then reversed during daytime, when melatonin production is switched off. Although the sleep–wake cycle is the most obvious circadian rhythm, there are several others related to it, including variations in body temperature, insulin production, and various functions relating to the endocrine, kidney and sex organs. Through its influence on the brain, the pineal gland similarly regulates our sensation of hunger and thirst, as well as influencing our mood and sense of well-being.

To say that these cycles are switched off and on by the presence or absence of daylight is an oversimplification. Experiments have shown that circadian cycles still occur, even without the stimulus of daylight, but they slow down by approximately 1.1 hours in every 24. In other words, the action of daylight is to speed up (or entrain) the body's circadian rhythms so as to coincide with the 24-hour daily cycle. This is described as phase shifting, and a positive phase shift of on average 1.1 hours is thus needed daily.

The Kronauer model

The intensity and duration of illuminance required to achieve this vary during the day. The pineal gland is much more sensitive early in the morning than during the middle of the day, as is shown in Figure 10.16. At peak sensitivity (at about 4 am), even low-intensity light can result in a positive phase shift, but as the day progresses, this sensitivity gradually declines, so that higher and higher light intensity (and/or longer duration) is required to have the same effect. According to the currently accepted model of Kronauer,[13] there is a crossover point (at about 2 pm), after which exposure to daylight increasingly slows circadian rhythms down again (i.e. resulting in negative phase shift). Although this may at first seem strange, it is Nature's way of ensuring that a balance of exposure can be obtained, despite the wide differences in daylight intensities according to season and latitude. Thus, for example, morning exposure to the very strong light of the tropics, which would otherwise shorten circadian cycles by far more than 1.1 hours, can be counterbalanced to the required extent by similar exposure in the afternoon. At the same time the system is sensitive enough to allow positive phase shifting, even on dull days at higher latitudes.

However, with increasing latitude, achieving sufficient exposure to daylight in winter can become problematic, particularly for the vast majority of the population who spend most of the day inside buildings. As can be seen from Figure 10.17, daylight (which we have evolved to

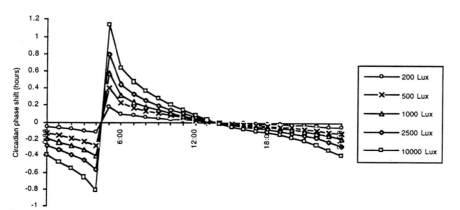

Figure 10.16 Kronauer's model of human sensitivity to light exposure in relation to phase shifting (shortening/lengthening) of circadian cycles.[12]

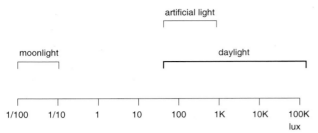

Figure 10.17 Typical illuminance ranges of different light environments.

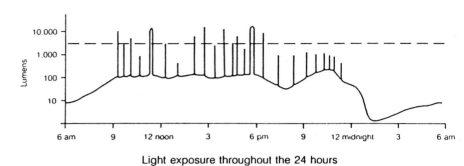

Figure 10.18 Light exposure over 24 hours working in an office in San Diego.[15]

respond to) is typically several orders of magnitude brighter than artificial light, even in cloudy conditions.

The low levels of exposure that are increasingly the norm in modern life are well demonstrated in Figure 10.18, where light detectors were worn by volunteers for stretches of 24 hours or more. It can be seen that exposure rarely exceeded 1000 lux and for most of the time averaged about 100 lux. This was despite the fact that the study was carried out in San Diego, one of the sunniest places in the USA. Many of us, even at low latitudes, are spending most of our time at light levels equivalent to twilight.

The health effects of circadian dysfunction

If the intensity of light reaching the retina is too low, the implications are far-reaching. Over time, the pineal gland's cycle of melatonin production will gradually lengthen towards its free-running value of 25.1 hours, and circadian desynchronization will be experienced. Melatonin will be released at the wrong times of day, resulting in lethargy, drowsiness and a variety of other symptoms caused by the mistimed activity of those organs regulated by melatonin.

For certain individuals, this gives rise to a condition known as seasonal affective disorder (SAD). Most people living at high latitudes experience seasonal changes in mood or behaviour to some degree, but SAD sufferers experience them to the extent of being seriously debilitated in winter. Fortunately they can usually be cured by bright light therapy in the early morning (when the pineal is most sensitive), using artificial light of between 2000 and 12 000 lux according to proximity to source, duration of exposure and severity of symptoms.[14]

As well as the retina not receiving enough strong light, there is also the possibility that irregular, intermittent exposures of the sort depicted in Figure 10.18 could interfere with our circadian rhythms. Compared with the regular pattern of exposure that would be experienced if we lived a more outdoor life, this could make

it more difficult to achieve the correct net positive phase shift and to balance morning exposure (which shortens circadian cycles) with afternoon and evening exposure (which lengthens them). It is possible that such random, but constant, interference could give rise to a variety of subtle minor misfunctions.

The full extent to which chronic circadian dysfunction may give rise to ill health is still the subject of medical research, but it is likely at the very least to result in impaired mental and physical function. While those living and working at high latitudes in winter are most at risk, inadequately daylit buildings can put occupants at risk at any latitude, as Figure 10.18 indicates, even in summer. In this connection it is of interest that the most common symptom typically reported in studies of sick building syndrome (see below) is lethargy. Financially, the slightest reduction in employee performance results in enormous cumulative costs in terms of lost production. In terms of quality of life, the costs of circadian dysfunction are incalculable.

Design implications

There is thus a clear need to provide adequate levels of daylight in buildings in order to facilitate positive circadian phase shifting. In general terms, this need strongly supports the strategy of relatively shallow-plan building layouts, together with high ceilings and/or lightshelves in order to raise illuminance furthest from the window. However, architects and engineers should be aware of the implications of the asymmetry of the Kronauer model, whereby sensitivity to positive phase shifting occurs in the mornings only.

For most of the year, the greatest opportunity for positive phase shifting occurs at home when people get up in the mornings, so, by the time they have travelled to work, their sensitivity will have significantly decreased. Nonetheless, model simulations of the net phase shifting achieved by available light levels in alternative scenarios have demonstrated that such pre-work exposures are not in themselves sufficient, and that without adequate daylighting at the workplace, the correct net daily positive phase shift is difficult to achieve.[16] The duration of morning exposure at the workplace in effect makes up for the reduced sensitivity. At high latitudes, as winter approaches, people will be exposed to less and less daylight before arrival at work, and in mid-winter many will reach work in the dark. At these latitudes, morning daylight exposure in non-domestic buildings thus becomes even more critical.

The asymmetry of the phase-shifting cycle creates problems for building designers, in that the morning light from east- and south-facing windows is much to be preferred to the afternoon light from west- and north-facing windows. While it is common in passive designs for the north side of a building to be occupied by ancillary spaces, it is impractical to employ this tactic for the west side as well. Moreover, the desirability for as many occupants as possible to be close to east and south side windows imposes considerable constraints on layout planning and floor space use, if all occupants are to receive equal opportunities for adequate exposure.

In urban contexts, various strategies can be adopted to optimise exposure to morning daylight. Buildings on the east and south sides of an urban block, for example, can be made higher than those on the west and north. This has the double advantage of maximising the floor space most exposed to morning daylight, while also allowing for the possibility of morning light being reflected into the west- and north-facing floor space on the other side of the street. For this to be effective, the south and east facades of buildings would need to be of high reflectivity. Breaks in the east and south facades would allow morning light to penetrate more deeply into the urban block. Louvres or other shading devices can also play a role by helping to reduce negative phase shifts in the afternoons on west-facing facades (although they need to be fully open in the mornings). Computer models are becoming available as design tools for optimising the phase shift characteristics of proposed designs, including improving the performance of existing buildings.[17]

Other health benefits of proximity to windows
Finally, it needs to be emphasised that exposure to morning strong light is not the only health benefit from proximity to windows. Several studies have indicated that access to views is also important for occupants' health, especially if the view is of trees and parkland as well as clouds and sky and other natural features. However, even if it includes only the latter, a view through an adjacent window enables occupants to keep in touch with external events. In this way, consciously or subconsciously, they are made continually aware of their social context, as well as receiving the complex stimuli relating to ambient conditions that throughout evolution have influenced human physiology and behaviour.

The significance of proximity to a window is shown in Figure 10.19 in a survey of six buildings during an investigation of sick building syndrome (SBS). The horizontal axis shows the proportion of occupants in a building answering affirmatively to the question 'Do you sit next to a window?', while the vertical axis shows the SBS score for each building in terms of occupant dissatisfaction. A high dissatisfaction score indicates a high number of SBS symptoms – lethargy, impairment of concentration, itchy eyes, stuffy or runny nose and other respiratory complaints – symptoms that typically disappear on leaving the building and affect at least 20% of the workforce. Although this survey included only six buildings, the strong correlation obtained is suggestive of the combined physiological and psychological value to human health of proximity to windows.

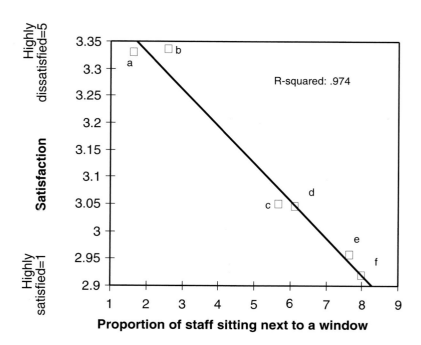

Figure 10.19 Proximity to windows and occupant dissatisfaction with the building.[18]

10.2.2 Exposure to ultraviolet light

Ultraviolet light is that part of the solar spectrum with a wavelength shorter than the visible waveband. Having higher frequency, it has more energy and for this reason can have damaging effects on biological organisms. However, the Earth's atmosphere scatters and absorbs a great deal of it, and in practice the intensity of light that humans are exposed to falls off rapidly below about 380 nm (the approximate limit of visible light), and very little UV reaches the Earth's surface below about 280 nm. That part of the spectrum from 380 down to 320 nm is commonly designated as UVA, and that from 320 down to 280 nm is designated UVB. In general, both the most harmful and the most beneficial effects of UV are the result of exposure to UVB. However, little UVB is transmitted through window glass.

In common with the rest of the solar waveband, the intensity of UV radiation varies according to latitude, season, time of day and extent of cloud cover.

The harmful effects of UV

Overexposure to UV radiation can cause damage to eyes (which cannot detect it), as well as to skin cells, giving rise to reddening (i.e. sunburn), skin 'ageing' and, ultimately, skin cancer. The most common form of skin cancer is, fortunately, rarely fatal and is nearly always curable. It is not malignant, in that it does not to spread to other organs in the body and tends to occur in old age, typically in fair-skinned people who have spent many years working outdoors or living in the tropics.[19] But there is one form of skin cancer, malignant melanoma, that does spread to other organs and is thus far more likely to be fatal. Fortunately, this form is rare (about 2% of all malignant cancers), although its incidence is rising in several countries (as is the incidence of the common form) for reasons that are not entirely clear.

The extent to which malignant melanoma is caused by exposure to UV radiation is still not known. In so far as UV exposure is implicated, it is currently thought to affect mainly those with very easily burnt skin that does not tan (i.e. the Celtic skin type), and only then if they have more than once 'flash fried' themselves, i.e. subjected previously unexposed skin to prolonged exposure to midday summer sun (e.g. on holiday or at an all-day sporting event). The risk appears to be greatest if this occurs in youth. In contrast, the link between UV exposure and common skin cancer is much clearer and is typically the result of a lifetime of cumulative exposure to sunlight. However, lifestyle factors (such as increased consumption of dietary fats) may possibly play a contributory part in the rising incidence. While some malignant melanomas may be the most dramatic manifestation of what one might call 'sun abuse', affecting comparatively few people, modern work and holidaying habits may be increasing the risk of skin cancer for a wider population. Without an intervening period of acclimatisation and the gradual acquisition of a protective sun tan, there is likely to be a penalty to pay for each year going straight from a dim indoor life to maximum exposure to the bright summer midday sun.

For all forms of skin cancer, the UVB component of ultraviolet light is several orders of magnitude more dangerous than UVA, although the latter is not entirely without effect. Following the discovery that the ozone layer is being depleted, there is current concern that the level of UVB radiation reaching the Earth's surface will significantly increase. For malignant melanoma, it is thought that this will be of little relevance compared with lifestyle factors.[20] But for the common form of skin cancer the cumulative burden of exposure over a lifetime will clearly be greater, at least for those who spend a high proportion of their time outdoors. However, since window glass transmits only a tiny proportion of UVB, this particular aspect of global climate change has few implications for building design. Some have suggested that more shading and covered walkways should be provided in outdoor spaces, but since most people now spend only a tiny proportion of their time outdoors, and tend to be adequately protected by clothing when they do, this is unlikely to have much impact on cumulative lifetime exposure. On the other hand, it would be advisable for buildings such as outdoor sports stadia to provide the option of shaded areas, especially if occupants are otherwise likely to spend long periods in the midday sun with less than normal clothing.

The beneficial effects of UV

While the dangers of overexposure to sunlight have received wide publicity, the dangers of underexposure are often overlooked. Human evolution has occurred in the context of exposure to daylight, including UV, and although the depletion of the ozone layer is a cause for concern, it does not alter the fact that human physiology depends on a certain degree of exposure to UV.

Chief amongst the benefits is the essential role that UV irradiation of the skin plays in the body's production of vitamin D, without which calcium cannot be properly absorbed and utilised. In the absence of adequate exposure to daylight, vitamin D cannot otherwise be synthesised by the body, and there is normally little in food. Studies have shown that in low-latitude countries both schoolchildren and, especially, pensioners tend to have significantly less vitamin D in their blood in winter than in summer. For children, low vitamin D can result in rickets, or bone deformation, while for pensioners the risk is for bone brittleness (osteoporosis), in both cases due to lack of calcium. Pregnant and lactating mothers are also at risk.

The amount of exposure to daylight required to achieve adequate vitamin D synthesis is difficult to specify owing to the wide variations in UV intensity. In the UK, for example, UV of the required intensity is available only from about 10 am to 4 pm from April to September. However, vitamin D is a soluble fat that is easily stored in the body, so that winter levels are the result of synthesis during the previous summer. Thus the higher the latitude, the more important it becomes to achieve sufficient synthesis during summer. In this connection, it is worth noting that the total effect relates to both the area of skin exposed and the duration of exposure. Although cloudy skies reduce UV intensity, they do not eliminate it; overcast skies typically reduce intensity by about 50%.

As well as promoting vitamin D synthesis, the action of UV light on skin may have several other benefits, but unfortunately these have not yet been adequately researched. Undoubtedly overexposure, especially to UVB, causes harm, but there is at least some evidence that moderate exposure is beneficial.[21] Reported effects include initiating defences against microbes and stimulating the immune system; increasing the oxygen-carrying capacity of the blood circulating in the capillaries close to the skin's surface, thereby reducing the work required of heart and lungs; and normalising both cholesterol levels and blood pressure, thus protecting against coronary artery disease. Russian studies have also shown that our ability to eliminate chemical contaminants is proportional to UV exposure. There is even some evidence that our skin has photoreactive enzymes that repair damaged DNA immediately upon exposure to sunlight.

Indeed it is possible that long-term exposure to sunlight has a protective effect against other forms of cancer. Those living near the equator suffer much less cancer in general than those living at higher latitudes. Cancer death rates for urban and rural areas in the USA have been shown to be inversely proportional to the solar radiation received. In a classic 1936 study, it was report-ed that while Navy personnel had eight times the national average for skin cancer, they had only 40% of the overall cancer death rate. Broad statistics of this kind are open to interpretation and are not necessarily conclusive, but they certainly give some credence to the widespread intuitive belief that sunlight is an important factor in promoting both good health and a sense of well-being.

The implications for window design
The relative importance of UVA and UVB in activating such beneficial effects is unfortunately known in only a few cases. For vitamin D synthesis, for example, it is thought that UVB is more effective than UVA, the latter making only a small contribution. Since window glass effectively excludes UVB, building occupants are thus cut off from most of at least one important beneficial effect of UV, although at the same time protected from the most harmful effects. Nonetheless, there is still the possibility, yet to be determined, that occupants may benefit from moderate exposure to the UVA waveband.

The proportion of UV transmitted by window glass varies with its type, thickness and angle of incidence. A single pane of ordinary clear 4 mm window glass, for example, transmits about 50% of the total UV radiation at 0° angle of incidence, averaging about 80% above 350 nm, but then falling rapidly to about 7% at 320 nm, i.e. all but a tiny portion being UVA. For most of the UVA waveband (320–380 nm), these transmission values remain relatively constant for angles of incidence up to about 60°. Above 60°, representing high solar altitudes in relation to vertical glass, they reduce quite rapidly. At UK latitudes and higher, this means that the UVA component of sunlight is largely unaffected by the angle of incidence, even in mid-summer. At lower latitudes, the angle of incidence will increasingly reduce transmissions, but mainly by capping peak mid-day intensities in summer. However, that part of the UVA waveband nearest to UVB, below about 350 nm, not only has lower transmission values, but also is more affected by the angle of incidence. This may possibly diminish the beneficial action of the net UVA that is transmitted through ordinary vertical windows. Moreover, UV transmission values for double glazing are significantly lower than for single glazing.

In many buildings, therefore, for most of the year the net UVA transmitted is unlikely to be of sufficient intensity to activate a physiological response, particularly at high latitudes. However, there is a lack of research in this area, and the possibility of achieving a benefit from UVA exposure within buildings (e.g.

through rooflights and inclined glazing) should not yet be ruled out. Since UV light has an unequivocally beneficial effect as a biocide against bacteria in the air and on exposed building surfaces, there are supporting arguments in favour of admitting it. On the other hand, UV light has other effects that may be less desirable, such as fading the colours of materials, paints and dyes. For this reason, glass manufacturers market a range of glazing products, for use in museums for example, that have much reduced UV transmission. This can be achieved by using either tinted or laminated glass (where the central plastic film absorbs almost all UV wavelengths). Increasing the iron oxide content of glass also reduces its UV transmission.

For general applications, however, one should perhaps beware of eliminating further segments of the solar waveband that we have evolved to respond to. It is interesting in this respect that a common feature of buildings implicated in sick building syndrome is the use of windows with tinted glass.[22]

10.2.3 General conclusions

When it comes to daylight and health, Le Corbusier's image of the sun is particularly apt (Figure 10.20). Its harmful and beneficial effects are inextricably bound up together, and it is difficult to avoid at least some risk of the former if one is to experience the latter. Clearly it is a question of balance.

Although there are some interesting questions relating to UV exposure, the principal daylight and health problem to consider relates to circadian cycles. Increasingly, occupants rely on buildings, certainly at high latitudes but also at low latitudes, to provide sufficient morning expo-

Figure 10.20 The two faces of the sun, by Le Corbusier.

sure to strong light in winter in order to achieve the correct positive phase shifting that they need. Apart from the passive solutions discussed above, there are of course technological solutions, involving artificial lighting. However, the energy penalties of providing the intensity of illumination required are considerable. One possible approach is to provide individual task lights at head height, perhaps coupled with controls to reduce the level of illumination gradually from an early morning peak. The short distance from source to retina means that much lower-strength lamps would be required than would otherwise be the case.

Artificial light fittings, however, bring with them problems of their own. The flicker of fluorescent lights is known to cause involuntary rapid eye movements that result in strain and fatigue (high frequency ballast can largely remove this problem). Studies have indicated that differences in the wavelength distribution of artificial lighting vis-à-vis daylight can also adversely affect occupants' mental and physical performance, thus exacerbating the effects of circadian dysfunction. Furthermore, it so happens that fluorescent lighting tends to have a dip in intensity in exactly the wavelength range (450–560 nm) that most activates the pineal gland.

These and other problems have prompted several commentators to suggest the use of full spectrum lighting in buildings, that is to say lighting that more closely mimics the wavelength distribution of daylight, especially in relation to UV wavelengths. This would bring indoors many of the potential benefits of UV exposure, including vitamin D synthesis, and indeed many studies have indicated that full spectrum lighting can be instrumental in curing or alleviating a wide range of medical conditions (Downing 1988). However, there is also concern that too many of the harmful effects may also be brought into buildings, and until these concerns have been more comprehensively addressed, full spectrum lighting is likely to remain unproven: a potential solution waiting for widespread application.

In the meantime, the provision of adequate daylight in buildings remains paramount on grounds of health, employing such passive means as shallow plan layouts, high ceilings and appropriate glazing ratios. In spite of all the advances in electric lighting technology, there still seems to be a strong human preference for being close to a window, and for buildings that are well daylit. This perhaps reflects an intuitively felt need and an essentially healthy impulse not to stray too far from the conditions in which we have evolved.

Summary

Visual comfort has both technical and psychological aspects relating broadly to visual performance and visual quality respectively. This is manifest in the actual ability to see fine detail and avoid disabling glare on the one hand, and the perception of pleasantness and enjoyment of the visual environment on the other. Whilst these two aspects can be initially treated separately, as with other environmental comfort issues, there is increasing evidence that they are often related.

Visual performance is dependent upon the apparent size of the object, the illuminance level, and the fatigue state of the observer. The contrast of the object relative to the background, and the directionality of the illumination are also highly influential. The presence of glare sources within the field of view, e.g. light sources or reflections, will also seriously impair visual performance.

Various proposals have been made to quantify glare in a lit interior; the glare index takes account of the size of the glare source, its brightness relative to the background, and the probability that it will fall in the normal field of view.

It is now generally accepted that the presence of daylight in buildings has a positive effect on the health and well-being of the occupants. It is suggested that this is partly due to the synchronising effect of high illuminance levels on the body's circadian rhythm; absence of regular exposure leads to phase shifting (relative to the persons daily activity) and symptoms (known as seasonal affective disorder or SAD) somewhat similar to jet-lag.

There is growing awareness of the danger of exposure to ultraviolet light (UV), present in unfiltered daylight and particularly sunlight. This has focussed interest on providing shaded areas around buildings. However, normal glass filters out most of the part of the UV spectrum which has detrimental effects, and thus sunlight does not normally present problems indoors. On the other hand, UV has some beneficial effects, stimulating vitamin production and acting as a bactericide.

References

1. N V Baker, A Fanchiotti and K Steemers (eds), *Daylighting in Architecture: A European Reference Book* (London: James & James (Science Publishers) Ltd, 1993), p. 2.3.

2. M J Tovée, *An Introduction to the Visual System* (Cambridge: Cambridge University Press, 1996).

3. R G Hopkinson and J B Collins, *The Ergonomics of Lighting* (London: Macdonald & Co (Publishers) Ltd, 1970).

4. R G Hopkinson, *Architectural Physics: Lighting* (London: HMSO, 1963).

5. CIBSE, *Code for Interior Lighting* (London: Chartered Institution of Building Services Engineers, 1994), p. 204.

6. J-J Meyer, D Francioli and H Kerkhoven "A new model for the assessment of visual comfort at VDT workstations", *Proceedings of Advances in Occupational Ergonomics and Safety, 1996.* (Cincinnati, Ohio: International Society for Occupational Ergonomics and Safety, 1996).

7. P Chauvel, J B Collins, R Dogniaux and J Longmore, "Glare from windows: current views of the problem", *Lighting Research and Technology*, vol. 14, no. 1, 1982.

8. M Seidl, "Tageslicht in Innenräumen", in J Beckert, F P Mechel and H-O Lamprecht (eds), *Gesundes Wohnen* (Düsseldorf, Beton-Verlag, 1986).

9. J A Lynes, "The window as a communication channel", *Light and Lighting*, November–December, 1974.

10. M-C Moore-Ede, F M Sulzman and C C Fuller, *The Clocks That Time Us: Physiology of the Circadian Timing System* (New Haven, Harvard University Press, 1982).

11. D Downing, *Day Light Robbery: The Importance of Sunlight to Health* (London: Century Hutchinson, 1988).

12. D Cawthorne, "Daylighting and occupant health in buildings", PhD thesis, University of Cambridge, 1995, p. 140.

13. C A Czeisler, R E Kronauer, J S Allan, J F Duffy, M E Jewett, E N Brown and J M Ronda, "Bright light induction of strong (Type 0) resetting of the human circadian pacemaker", *Science*, vol. 244.

14. Downing, *Day Light Robbery*.

15. *Ibid*, p. 25.

16. Cawthorne, "Daylighting and occupant health in buildings".

17. *Ibid*.

18. Building Use Studies, London.

19. IARC, *Solar and Ultraviolet Radiation*, IARC Monographs on the Evaluation of Carcinogenic Risks to Humans, no. 55 (Lyon: World Health Organisation International Agency for Research on Cancer, 1982).

20. A J McMichael, A Haines, R Sloof and S Kovats (eds), *Climate Change and Human Health: An Assessment Prepared by a Task Group on Behalf of the World Health Organisation, the World Meteorological Organisation and the United*

Nations Environment Programme (Geneva: WHO, 1996).

21. Downing, *Day Light Robbery*.

22. J M Sykes, "Sick building syndrome", *Building Services Engineering Research and Technology*, vol. 10, no. 1.

Part 3
Design criteria and data

11 Design tools

Design tools

The definition of a design tool offered in the literature is:

"A method, technique, procedure, equipment etc. that aids in the evaluation of a design, in making the 'correct' choices, often towards specific targets, i.e. compliance with codes, standards etc."

This very broad definition makes almost every kind of tool used in the design process (except for the pencil!) a 'design tool'. Many objectives are suggested for the use of design tools. Most of them fall into these categories:

- predicting daylight levels from diffuse skylight in rooms
- analysing solar access to rooms and surroundings of buildings
- calculating glare from daylight sources
- predicting the performance of shading devices
- predicting the performance of innovative daylighting systems and lighting control systems
- predicting energy savings from daylighting
- analysing the cost-effectiveness of daylighting systems and lighting control systems
- visualising the proposed daylighting design

The general impression is that the majority of the tools available are intended to be used at late stages of building design, as both window schemes and surface reflectances are necessary inputs. Most methods are based on the principle of analysing a given physical solution; very few are 'inverted', giving as output a suggested solution to a specific design target.

Computer-based tools have naturally become more dominant in later years as the hardware has become more available and powerful. Another newer trend is the introduction of 'integrated' tools that may carry a daylighting analysis all the way to calculate energy, economy and comfort.

11.1 Manual tools

Manual design tools are defined here as not requiring a computer for application. They range from simple rules of thumb, via mathematical formulas, tables, nomograms etc, to more elaborate physical equipment designed for different tasks during the design process.

The literature gives description of a large range of such manual tools. A recent survey is given in *Daylighting in Architecture*.[1] The survey comprises 54 manual tools:

- 7 equations
- 5 single-stage methods
- 3 lumen methods
- 5 tables
- 10 nomograms
- 8 protractors
- 5 dot diagrams
- 6 Waldram diagrams
- 5 urban analysis methods

The surveys of simple tools and computer tools in this reference book were undertaken in 1992. One might expect quite a number of new tools to have been introduced since. One might also expect that there are many more tools available that have not been presented in English language references, making a comprehensive list quite large.

A common feature of many of these tools is that they are designed to calculate the daylight factor of a defined win-

Table 11.1: Manual design tools in use: survey by DLE.

Country	Manual design tools
Denmark	BRS Daylight Protractors
UK	BRS Daylight Protractors
France	CSTB abacuses (the abacuses provide sky components for CIE overcast sky)
Germany	Scale models
	Calculation of methods described within DIN 5034
Ireland	Daylight factor meter (Working in the City Competition)
	Cylindrical sunpath diagram
	Scale models
Norway	Löfberg: *Räkna med dagsljus*[a]
	SS 19 42 01: *Dagsljus - Förenklad metod för kontroll av erforderlig fönsterglasarea*[b]
Portugal	BRS Daylight Protractors
	Sunpath charts (stereographic and cylindrical projections)
Scotland	BRE Design Code
Spain	RAFIS (Rough Analysis For Illuminated Spaces)
Sweden	Löfberg: *Räkna med dagsljus*[a]
	SS 19 42 012: *Dagsljus - Förenklad metod för kontroll av erforderlig fönsterglasarea*[b]
	Glaumann: *Sol i bebyggelseplanering*[c]
Switzerland	DIN 5034 (based on Waldram diagrams)
	Scale models and sky simulator

[a] Räkna med dagsljus. Hans Allan Löfberg. Statens institut för byggnadsforskning. Gävle, Sweden 1987.
[b] Byggnadsutforming - Dagsljus - Förenklad metod för kontroll av erforderlig fönsterglasarea. Sweden 01-01-1988.
[c] Sol i bebyggelseplanering, Glaumann, M. Statens råd för byggnadsforskning, T 1976:37. Stockholm, Sweden, 1976.

dow scheme. Tools for the earlier phases of design are not included. Other surveys[2,3] give the same impression.

A more limited survey was conducted among the participants in the Daylight Europe project. The resulting information is listed in Table 11.1.

To what extend are such tools used, and which tools are preferred in the design community? *Daylighting in Architecture* also quotes a survey of the use of tools conducted among architects in south European countries, which showed that only about 15% were concerned with daylighting issues and were able to use design tools.

The most recently published study of design tool use has been undertaken by the Building Research Establishment.[4] This study also reviews earlier work. A questionnaire was sent to 323 designers; about 20% were returned. Of these, 30–55% never made predictions, the remainder infrequently depending on the problem at hand.

The respondents identified 18 computer programs and 27 manual methods. Computer programs were used mainly by daylighting specialists. More surprisingly, the simple average daylight factor formula was used by only 20% of the respondents, and many of the users of simple tools never calculated the internal reflected component, thus seriously understimating the daylight in the back of rooms.

The study concludes that there is a place for all sorts of daylight prediction methods – hand calculations, scale models and computer programs – but there is a need both for tool improvements and, in particular, for better dissemination and training.

We may conclude therefore that experts in the field of daylighting, i.e. tool developers, researchers, teachers, etc., probably overestimate the 'market' for such tools and the importance that designers place on daylight consideration in the design process.

11.2 Computer-based analytical tools

With the complexity involved in accurate daylight calculations, it has become evident that computers are the logical instrument. The first generation of computer-based design tools for daylighting were limited to simple geometry, in principle a rectangular shoe-box room with windows in one wall only. The calculations were based on algorithms for the direct sky component from rectangular windows and the BRS split-flux formulae for the interreflected component. Many of these tools offer graphic output in the form of illuminance contour maps or perspective drawings of the illuminance map. Most of the programs in this family were developed for the MS-DOS and Macintosh operating systems; they never reached widespread use.

This generation of tools have now been superseded by PC programs for the Windows environment. Another development has been the transfer of complex mainframe tools to the PC world. Input of geometric data is for some tools now linked to CAD programs, and some daylight tools can produce images based on the calculation results.

However, using computer-based daylighting analysis tools is still not a common feature of normal architectural design practice. The tools presented here have been selected on the criteria that they are known to be used in several European countries, and represent the state of art in their respective group.

LESO DIAL

This program for the Windows environment calculates daylight factors from analytical algorithms, for CIE overcast sky conditions.[5] It uses analytical expressions for the direct sky and outside reflected component, and the BRE split-flux formulae for the internal reflected component. In addition, it provides the user with some decision support in improving the tested design.

The program is easy to use, it requires very little tutoring, and includes a vocabulary of lighting terminology. The program is intended for early stages of design decision-making. It deals with both windows and skylights.

The user first defines the activity in the space to be analysed; this is used to set the required illuminance level. Geometrical input can be given by dragging and stretching lines on the computer screen. Reflectivities can be input as verbal statements; fuzzy logics handle the parameters not precisely known in numbers.

The program calculates daylight factors and daylight sufficiency, the latter being the percentage of time when electric lighting can be switched off. This feature uses Swiss climatic data; one disadvantage is the lack of a provision for input of user-defined daylight climate.

The program also has a diagnostics feature; it will provide comments on the results and suggestions for improvement of the analysed design. Another feature under development is a comparison module where the calculated case is compared with monitored case studies stored in a database. The database will be scanned to find a case that is similar to the calculated case.

A new version covering all European sites is shortly to be developed under the EU funded project, DIAL-EUROPE.

PASSPORT LIGHT

This program was developed as part of the European daylighting project.[6] The calculation procedure is backward ray tracing similar to an algorithm in the RADIANCE program (see below). It provides good freedom in geometry: it can handle up to 500 surface elements, which have to be rectangular. The surfaces can have five different characteristics as to reflectance distribution, but only monochromatic surfaces are modelled.

PASSPORT LIGHT can analyse CIE overcast and clear sky, uniform sky and user-defined sky luminances. The computing time is short, but input is difficult and time-consuming. A graphic output for geometry provides an input error check.

The program functions somewhat like a photocell in certain defined positions: it adds up contributions from all the surfaces visible from the point in question. The number of computing points is limited.

SUPERLIGHT

This program was one of the first tools widely available that provides accurate calculations through a finite element radiosity method.[7] The current version handles both daylighting and electrical lighting, the daylighting for CIE overcast, CIE clear, and uniform sky with and without sun. The daylight apertures can be both ordinary windows and skylights.

The SUPERLIGHT program allows quite complex room geometry: L-shaped rooms, interior partitions and external obstructions. A graphical display of input geometry is available for checking purposes.

Surfaces have to be perfectly diffusing; specular surfaces are not included. The glazing can be both clear and diffusing; windows can have curtains, but not venetian blinds.

The program provides output in tabular or graphical form as illumination or daylight factor mapping on the room plan. Results are available for all room surfaces and user-defined working planes. The output can be transferred to the RADIANCE program (see below) for visualisation.

ADELINE

ADELINE (Advanced Daylighting and Electric Lighting Integrated New Environment) is a program package that combines daylighting design tools with energy simulation tools.[8] The daylighting analysis is performed with either SUPERLIGHT (see above) or RADIANCE (see below) for a certain set of sun positions and sky conditions.

Hourly lighting energy use is then calculated, by comparing the interior daylighting levels with required lighting levels and control strategies. The daylight levels are found in the data set provided by the daylighting programs, using the actual sky condition and sun position for that hour as parameters. The resulting lighting energy schedule is then fed as input to an energy simulation program: the user has a choice of SUNCODE, DOE 2, TRNSYS or TSBI3.

The geometric input to this program package can be done by a special graphical interface (Scribe modeller). CAD files will also be accepted.

11.3 Computer-based image solution tools

RADIANCE

The RADIANCE program is a complete lighting visualisation system, as it provides both numerical results and photo-realistic rendering.[9] The program handles CIE overcast and clear sky, the latter with or without sun. The CIE distribution can be modified with different parameters, and the program also accepts user-defined skies. RADIANCE can also model electric lighting.

The input can be made via a CAD system that provides DXF files, and the input geometry can be checked in a graphical output facility. RADIANCE can model plane surfaces, spheres, cones, and combinations of these. A wide range of surface reflectance and transmittance properties are allowed: diffuse, specular, intermediate viewing-angle dependent anisotropic etc. Spectral data (i.e. colours) can be modelled.

The lighting analysis is based on a backward ray-tracing technique. Rays of light are generated from the point of measurement, and followed backwards through reflections until it reaches a light source or is extinguished.

The output can be tables, illumination or luminance mapping, or visualisation from any viewpoint. Since the program handles angle-dependent surface properties, changing the viewpoint will require a full new calculation; this is a major difference from radiosity calculations based on perfect diffusing surfaces only.

RADIANCE is a very accurate program with many powerful application possibilities. The program requires a fairly long training period before all the possibilities available are mastered, however.

GENELUX

GENELUX is the result of a European development, in some respects similar to RADIANCE in capabilities.[10] The program models various sun and sky conditions, also electric lighting. Surfaces can be specular, diffuse, or combinations, and input can be spectral for colour analysis.

The calculation procedure is based on forward ray tracing from the light sources to the measurement points. Rays hitting a surface are spread into many new rays through a Monte Carlo technique. An option is to use radiosity analysis for the diffuse portion of the surface reflectances.

Input of geometrical data can be via the CAD programs ArchiCad and AutoCad. Output includes calculation of DGI (daylight glare index).

GENELUX is also available in a Web version, Genelux-Web, which has some restrictions on the capabilities.

This program calculates interreflections in a space only above the user-defined working plane, which means that the floor cavity below a table is simulated through an equivalent reflection factor. This procedure will necessarily incur some inaccuracy.

The results of the computer design tools survey conducted among the participants in the Daylight Europe project are listed in Table 11.2.

Table 11.2: Computer design tools survey.

Country	Computer design tools
Denmark	ADELINE
	PROLIGHT
	TSBI3
England	RADIANCE
	ANGLIA DAYLIGHT
France	QUICKLITE
	GENELUX
Germany	ADELINE (Superlite, Radiance, Superlink, Radlink), for simple and fast studies as well as advanced daylighting design studies
Greece	RADIANCE
Ireland	HEVACOMP
	FACET
	WIS
	ANGLIA DAYLIGHT
Italy	KANDALA, developed by Conphoebus s.c.r.l., Catania
	LIGHTPROJECT, developed by Architecture Department of University of Venice
Norway	ADELINE (Superlite and Radiance)
Portugal	LUZNAT – PC-based software tool
	RADIANCE
	DOE-2; VisualDOE-2, ESP-r
Scotland	RADIANCE
Switzerland	ADELINE software package
	DIAL

A rigorous comparative study of ADELINE (RADIANCE), LIGHTSCAPE, MICROSTATION and RADIORAY has been carried out by Ashmore and Richens[11] recently. It also compares simulated results with those using two artificial skies, a mirror 'CIE type' sky and a 145 sector dome sky.

Physical models and artificial skies

11.4.1 Principles of physical modelling of daylight

Scale models of buildings are used for the purpose of daylighting design all around the world. Models may be illuminated under a real sky, or more commonly, for convenience and standardisation, under an artificial sky.

The main advantages of this approach, compared with other design methods, are as follows:

- Architects have used scale models for centuries as general design tools for studying various aspects of building design and construction.

- It is a 'soft technology', well mastered and shared by architects and other building professionals.

- When properly constructed, they portray the distribution of daylight within the model room almost exactly as in a full-size room. This is due to the extremely small size of light wavelengths (380 – 780 nm), compared with the size of even the smallest scale model. The physical behaviour of light is the same for a full-size room as for a scale model for all practical scales.

11.4.2 Main construction rules for scale models

Model construction must be preceded by the choice of an appropriate scale, which is directly related to the particular purpose. Table 11.3 summarises the possible choices.

Note that common architectural scales follow the series 1:10, 1:20, 1:50, 1:100 etc. Moving from the scale 1:50 to 1:20 results in a volumetric increase of almost 16 times. This may result in inconveniently large or small models being used. To avoid this it is possible to work at an intermediate scale, using the zoom function on a photocopier to change scales.

Common rules must be applied, however, in the construction of the model, whatever the scale. The principal rules applying to cases where conditions are assessed inside the model, are as follows:

Table 11.3: Scale choice in relation to purpose.

Scale	Purpose
1:500–1:100	For preliminary design and concept development
	To provide an overall sense of the massing of the project
	To study the shadows cast by the future building or by neighbouring buildings
1:200–1:10	To determine unwanted reflections on a glass facade
	To study direct sunlight penetration into a building (e.g. efficiency of solar protection)
	To study diffuse daylight in a very large space (e.g. atria)
1:100–1:10	To consider detailed refinement of spatial components
	To obtain highly detailed inside views (e.g. for video or photographs)
	To study accurate diffuse and direct daylight penetration

- The walls of the model should be absolutely opaque.

- All the joints should be lightproof.

- Where appropriate, model parts should be movable or replaceable to facilitate comparison of configurations.

- Reflection coefficients of internal (walls, ceiling and floor) and external surfaces (ground, obstructions, etc) should be as close as possible to those of the proposed building.

- Geometric accuracy should be as high as possible, especially for apertures, window frames and glazing.

- If it is known that in use the room will be heavily obstructed by furniture, particularly if above the workplane, then this should be modelled – crude models are quite sufficient provided they are of appropriate reflectance.

- If there are obstructions close to the site being modelled, these must also be modelled, although often they can be simple 2D shapes.

- When using unglazed models transmission factors have to be applied to allow for glazing – models

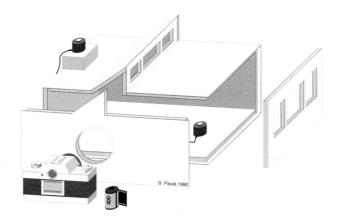

Figure 11.1 Example of a scale model used for a daylight simulation.

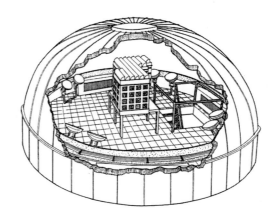

Figure 11.3 Lawrence Berkeley Laboratory, USA. View of a typical sky dome.

where light enters the apertures very obliquely (e.g. buildings with atria or lightwells) should be glazed to avoid errors caused by the strong angular dependence of the light transmission of glass.

- The overall dimensions and weight of the model should be such that it can be supported easily (e.g. on a heliodon).

- The size of the model should be small in relation to the distance to the light source (e.g. 0.6 m in height for a 5 m diameter sky dome).

- Access inside the model, through apertures or removable parts, should be possible for placing illuminance sensors or imaging devices.

Figures 11.1 and 11.2 illustrate some of the most relevant construction rules. Because of the difficulty in observing all of these constraints, physical modelling generally achieves relative, rather than absolute, results. The search for relative improvements in performance is thus a more appropriate goal than attempting to obtain absolute quantitative measurements.

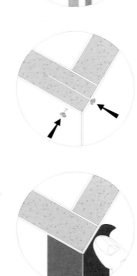

Figure 11.2 Detail of construction of scale model.

11.4.3 Sky simulators

Sky simulators have been used for many decades for daylighting design studies. Their main advantage is to offer reliable and reproducible conditions that simulate external daylighting. The use of normalised sky luminance distributions, so called 'standard skies', makes it possible to compare daylighting design studies carried out with different simulators.

Table 11.4 gives an overview of the principal sky simulator configurations. One of the earliest sky simulators, the sky dome, already reported in the 1930s,[12] is shown in Figure 11.3.

Figure 11.4 illustrates the most common (and lowest cost) configuration, the mirror sky. It consists of a dif-

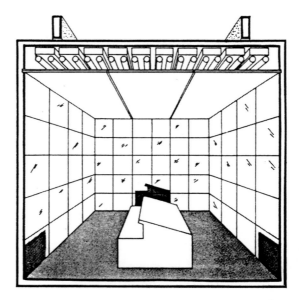

Figure 11.4 Centre scientifique des Techniques des Bâtiments, France. View of a mirror sky simulator.

Table 11.4: Principal sky simulator configurations.

	Description	Advantages	Disadvantages
Mirror sky	Most common configuration. Mirror enclosure with a lighting ceiling (fluorescent tubes and opal diffuser)[13]	Moderate cost Minimised horizon error	Only CIE overcast sky reproduced Inter-reflection disturbed by the scale model
Sky dome	Diameter between 3 and 9 m;[14] made of white opaque hemisphere illuminated by internal light sources in a circular groove or truslucent dome with external illumination[15]	Reproduction of different standard sky models (uniform sky overcast or clear CIE sky) possible Very easy scale model access	Hard and tiresome calibration (about 1 week) High electric consumption and frequent maintenance problem
Spotlight sky simulator	Vault made of a multitude of incandescent lamps[16]	All types of sky reproducible	Calibration and maintenance complicated by different ageing patterns of sources High luminance discontinuity and multiple shadows
	Line of 30 lamps mounted in a quarter-circle arc[17]	All types of sky reproducible Moderate cost	Model cannot be viewed under simulated daylight Slow measurements procedure
Scanning sky simulator	Only a sixth of the vault is constructed with 25 lamps. The whole hemisphere, according to the Tregenza's model (145 light zones), is rebuilt by a six steps scanning. Quantitative (illuminance) and qualitative (video digitised image) are add up in the end of the process.[18]	Has a close matching to the sky luminance measuring format Reproduces all existing standard or statistical sky models Achieves low construction, maintenance and operation cost	It is impossible to visualise or to measure instantaneously inside the model since images have to be reconstructed

fusely luminous ceiling – usually a translucent diffuser illuminated by banks of fluorescent tubes, and four walls of plane mirrors. Interreflections between the mirrors form an image of an infinite diffuse sky, which owing to absorption in the glass diminishes its brightness as the 'virtual horizon' is reached.

Several novel sky simulator designs have been proposed over the last few years (Table 11.4). They all aim to increase ease of use and to extend the possibilities of simulating different sky distributions. One of the most recent is presented below.

11.4.4 New sky simulator configurations

Some of the new sky simulator configurations being proposed are based on a scanning process.[19, 20] Of these, one in particular uses new video imaging capabilities to retrieve maximum quantitative information from scale model studies.

The device uses a scanning process to rebuild the overall sky hemisphere, starting with a part of it (a sixth of a hemisphere). This novel apparatus possesses some important advantages, which are summarised in Table 11.4.

Description of the scanning sky simulator
This new sky simulator, illustrated in Figure 11.5, is composed of:

- a luminous vault

- a rotating model support

- a control, monitoring and visualisation unit

Figure 11.5 View of the scanning sky simulator (the scale model is placed on a rotating support at the centre of the simulator). (LESO-PB/ITB/EPFL)

The adopted configuration of the luminous vault is based on the measuring format proposed by Tregenza[21] within the framework of the International Daylighting Measurement Programme (IDMP). It has the following advantages:

- The hemisphere is 'tiled' with identical discs, achieving a covering of 68% of the total surface.

- It is split vertically into six symmetrical parts, each a rotation of 60°.

- It corresponds to the IDMP standard sky luminance measurements, so that the direct use of these data is made possible.

Figure 11.6 illustrates this configuration and indicates at the same time the dimension and position of the luminous discs that constitute the sky vault.

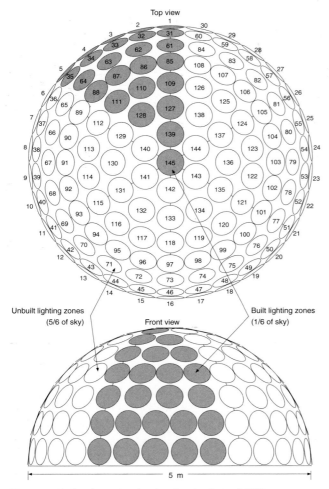

Figure 11.6 Configuration for the sky simulator. (LESO-PB/ITB/EPFL)

In order to limit the problems of maintenance and calibration of the apparatus, which arise owing to the differential ageing of lamps, and to reduce its energy consumption, cost and size, only one sixth of the sky was constructed.

This was possible by taking advantage of the vertical axis of symmetry of this configuration. The overall hemisphere is reproduced by means of six successive rotations of the scale model under the simulator, 60° at a time.

The luminous vault that reproduces the characteristics of the reduced sky vault consists of 25 luminous discs spread over the one-sixth hemisphere (of 5 m diameter) in the same configuration as described by Tregenza for the whole hemisphere. These luminaires are held in place by a structure of metal tubes.

The contribution of a luminaire to the illumination of a model at a distance of 2.5 m amounts to 140 lux. The horizontal illumination from the 25 luminaires is about 1700 lux (isotropic sky); the equivalent illumination of the whole vault is about 10 200 lux.

Measurement principles

The evaluation of the luminous performance of a daylighting design should ideally include:

- quantitative studies based on the use of photometers to measure daylight factors

- qualitative studies based on the use of a video system to visualise subjective qualities and to evaluate visual comfort.

However, a scanning sky simulator does not allow either a direct view or direct measurements, and requires the following procedure:

- successive rotations of the scale model coordinated with an appropriate distribution of the source intensities (possibly changed at every rotation)

- recording of illuminances with photometers and digital images with video at every rotation

- daylight factor calculation and visualisation of the resulting images after reconstruction

The whole procedure is automatic and takes less than 2 minutes. A personal computer is responsible for controlling the rotating support, acquiring photometric measurements, digitising the video images from a high-resolution camera, and the processing of all the acquired data.

This procedure can be completed by visual comfort analysis. Initial calibration of the camera is necessary, as well as processing the images by the appropriate computer program.[22, 23] Figure 11.7 illustrates this procedure in the case of a simple daylit room.

Figure 11.7 View inside a scale model (daylit room): final image made of the addition of the six partial images, associated to the six subdivisions of the sky vault. (LESO-PB/ITB/EPFL)

Sky simulation possibilities

The novel principle of the scanning sky simulator allows an accurate reproduction of the luminance distributions of every type of sky. Some of these distributions, standardised by the CIE recommendations, are described by mathematical equations. These distributions are used in daylighting studies and, although theoretical, they have the important advantage of allowing results to be compared internationally.[24] The main 'standard skies' that can be reproduced on the simulator are:

- isotropic overcast sky
- CIE overcast sky[25]
- CIE clear sky[26]
- CIE intermediate sky[27]

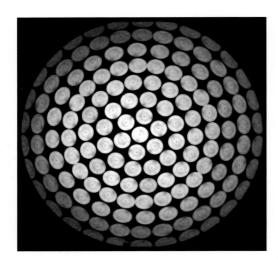

Figure 11.8 Fish-eye view of a simulation of an overcast CIE sky model (Moon and Spencer's model). (LESO-PB/ITB/EPFL)

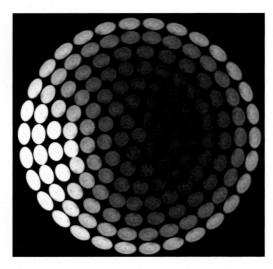

Figure 11.9 Fish-eye view of a simulation of a clear CIE sky model (Kittler's model, country, sun altitude = 30°). (LESO-PB/ITB/EPFL)

Figures 11.8 and 11.9 give a view of these types of sky, simulated by the device. Besides the different standard skies, moreover, it is possible to reproduce statistical skies. Through the IDMP, illuminance and luminance distribution of real sky measurements have been made available.[28] The processing of these data allows the development of statistical skies that are representative of the daylight in a particular area. Monthly average skies, as well as dynamic daily skies, can be reproduced in this way.

11.4.5 Application case study

The daylighting of an administrative office in Lausanne, Switzerland, is used to illustrate an application case study.

The preliminary daylighting concept of the building was to use a lightshelf in order to offer solar protection, as well as a better distribution of daylight in the offices. The following parameters of the lightshelf were analysed with the scanning sky simulator:

- slope and coating of the lightshelves
- glass type (clear or diffuse)

Performance was compared with and, the reference case, without the lightshelf.

Results of daylighting design

Figure 11.10 summarises the main outcome of the design, which led to the following conclusions:

- In the −10° to 5° range, the slope of the lightshelf has little influence, with either a diffuse or a specular (glossy) coating (slopes are needed to allow rain to drain off).

- In the back of the room (3–5 m away from windows), there is 25% more light with a white than with a dark lightshelf.

- For the same reflection coefficient, a diffuse coating is better than a glossy one.

- Diffuse glass has two disadvantages compared with clear glass: it increases light in the front of the room, and it has a lower transmission coefficient.

Compared with the reference case, the lightshelf leads to a decrease in the total diffuse daylight flux into the room, but a more uniform distribution, thereby reducing the eye's adaptation stress.

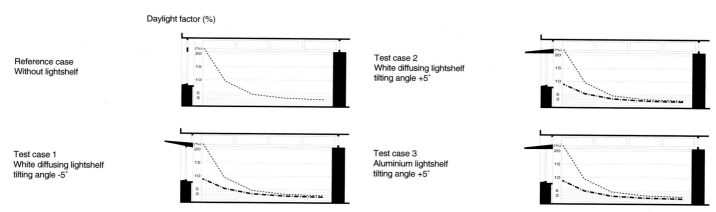

Figure 11.10 Summary of the main outcome of the design of an administrative building located in Lausanne. (LESO-PB/ITB/EPFL)

Practical DF measurement procedures for fixed skies

11.5.1 Basic measurements

For model testing in non-scanning skies – i.e. mirror skies or diffuse dome skies – the following proceedures can be applied in most cases. It is assumed that an illuminance meter is available with one or more sensors, small enough to be placed in the model at the correct scale height. The sensors are usually about 1 cm high, and should be cosine corrected.

1 Switch on lighting and allow 1/2 hour to stabilise – particularly necessary for fluorescent lighting.

2 Assuming there is a reference cell plus one or more measuring cells, place cells next to one another on horizontal surface on top of model. Make sure that your body is below the 'horizon' to avoid overshadowing the cells.

3 Record illuminance values for measurement cells and reference cell. Ideally these will all be the same, but if significant differences are found (more than 5% error) then calibration factors can be calculated for each measuring cell.

4 Leaving the reference cell on top of the model, move the measuring cells into model and place cells with their centres over the points to be measured. For most applications the sensitive surface of the cell should be at the scale workplane height – 900 mm.

Usually a series of measurements is made in order to plot a profile or contour map of the daylight factor. If there are a number of measuring cells available they can be set out as an array in the model. If there is only one measuring cell, it will have to be repositioned between each measurement.

Ensure that the cell/cells are positioned correctly and horizontally, and that you do not obstruct the sky from the model.

If there is only one cell, it becomes even more important to allow the lighting system to stabilise. A reference reading is taken before placing the cell in the model and then further reference readings at intervals as necessary, according to the stability of the sky illuminance.

11.5.2 Calculation

If the cells do not require a calibration factor then the DF at any point n is given by

$$DF = 100 \times GF \times E_n / E_{ref}$$

where GF is the glazing transmission factor described below, E_n is the illuminance at point n and E_{ref} is the reference illuminance.

If calibration factors (C) are necessary then E_n should be multiplied by C_n where

$C_n = E_{ref} / E_{n\ cal}$

and $E_{n\ cal}$ is the calibration reading taken when next to the reference cell.

If the model is not glazed, the DF has to be corrected by the transmission factor GF as given below:

- single glazing   0.85
- double glazing   0.75
- triple glazing   0.70

If glazing bars have not been modelled physically, then an allowance must be made for this. Typical framing layouts obstruct the aperture more than one might expect (5.1.3); up to 40% is not uncommon.

A further correction factor has to be applied to take account of the dirtying of the glass (Table 11.5). For more detailed values of other glazing types consult the light transmission data in section 12.3.

Table 11.5: Correction factors to allow for dirt on glazing.

Location of building	Inclination of glazing	Type of work in building	
		Clean	Dirty
Non-industrial area	vertical	0.9	0.8
	sloping	0.8	0.7
	horizontal	0.7	0.6
Dirty industrial area	vertical	0.8	0.7
	sloping	0.7	0.6
	horizontal	0.6	0.5

11.5.3 Effect of orientation

The effect of orientation on the availability of daylight has already been discussed in section 3.2.1. Orientation factors for the five European daylight zones are given in section 12.2.

Normally a single orientation factor is applied at the daylight sufficiency calculation stage, corresponding to the orientation of the principle glazing. If the room being tested has glazing in more than one orientation then a more accurate method is to calculate an orientation-weighted DF. This is carried out as follows:

1 Obstruct all apertures in the model except for one orientation. Measure, and calculate the DF for the particular point.

2 Multiply this DF by the appropriate orientation factor.

3 Carry out the same procedure for all orientations.

4 Add the corrected DF contributions to give the total DF.

Obviously this should be done for every measurement position since the relative impact of individual apertures will vary across the room. It may be justified only where high precision is required but, as can be seen from the magnitude of the orientation factors, failure to carry out this procedure could lead to errors of up to 50% in sizing individual apertures.

Note that the procedure has the effect of converting the cylindrically symetric sky into a more realistic asymetric sky.

Caution must be exercised not to apply the correction twice. Firstly, if the artificial sky already has a non-cylindrical distribution then the orientation factors are not required at all. Secondly, if they are applied at this stage to evaluate an orientation-sensitive DF, then they must not be re-applied when calculating daylight sufficiency as described in 12.2.

11.6 The BRE daylight factor protractors and IRC nomogram

11.6.1 The sky component (SC) and the externally reflected component (ERC)

Four of the protractors out of a full set of ten are reproduced in Figures 11.11 to 11.14. They may be photocopied onto acetate; alternatively room plans and sections on tracing paper may be placed over them. It is recommended to use the CIE Overcast distribution for mid and northern Europe (protractors 2 or 4), and the Uniform Sky distribution for southern Europe. Take the following steps to establish the SC with reference to Figure 4.38.

1   On a section of the room through the main glazing area, draw the workplane and on it the point to be considered (O).

2   Connect the limits of the aperture (or edges of an external obstruction such as a wall or an overhang) to point O, i.e. the lines PO and RO.

3   Place the protractor (with sky component scale uppermost) with centre at point O and baseline on working plane.

4   Read values where lines PO and RO intersect the outer scale; the difference between these two values is the initial sky component SC_i.

5   Read the elevation angles where lines PO and PR intersect the inner scale and take the average of the two readings – this is the elevation of the aperture.

6   Take the room plan and mark the position of the point considered at O.

7   Connect the limits of the aperture to point O, i.e. the lines MO and NO.

8   Place the protractor with the correction factor scale towards the window, the baseline parallel to the window, and the centre at point O.

9   Mark four concentric semicircles on the protractor 0°, 30°, 60° and 90°. Select one according to the elevation angle measured in step 5, interpolating if necessary an imaginary semicircle.

10 From where lines MO and NO intersect this semicircle drop down parallel to the curved 'droop lines' to the inner circular scale and read off the values.

11 If the two intersection points are on either side of the centre-line, as in the diagram, add the two values; if they are on the same side, subtract the smaller from the larger. The value obtained is the correction factor.

12 Multiply the initial SC (step 4) by the correction factor to obtain the final SC.

If there are no obstructions outside the window there will be no ERC. If however there are objects higher than the line RO, the light reflected from these objects will reach the point O and contribute to the illumination at that point. This contribution is the ERC and can be found as follows:

1   Calculate the SC_{obs} for the part of the sky covered by the obstruction, following the steps described above.

2   Multiply this value by the luminance of this surface relative to the sky, which is calulated to be: (reflectance of the surface) x 0.5. If the reflectance is unknown it is customary to assume this to be 20%. This is the externally reflected component (ERC).

11.6.2 Internally reflected component

Much of the light entering the room will reach point O after reflection from the walls, ceiling and other surfaces in the room. The magnitude of this contribution to the illumination at point O is expessed as the internally reflected component (IRC). This will normally be fairly uniform throughout the room: thus for most cases it is sufficient to determine the average IRC value. The simplest method uses the nomogram in Figure 11.15. Steps to be taken are as follows:

1   Find the window area and the total room surface area (floor, ceiling and walls, including the windows) and calculate the ratio of window area to total surface area. Locate this on Scale A on the nomogram.

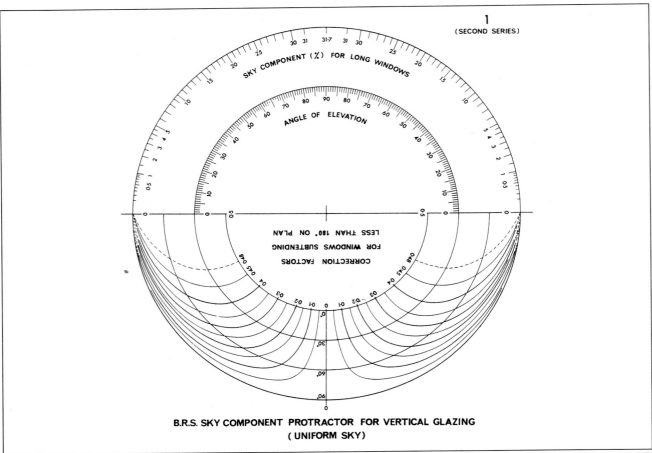

Figure 11.11 Sky Component Protractor No.1: Vertical glazing, Uniform Sky (BRE Copyright, reproduced with permission. Full set available from CRC, 151 Rosebery Avenue, London EC1R 4GB)

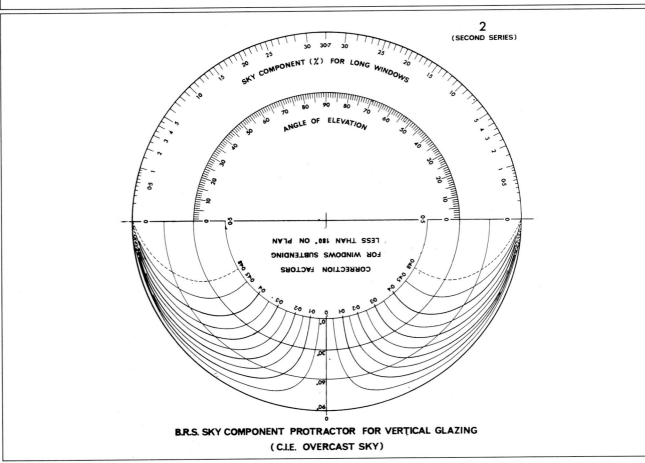

Figure 11.12 Sky Component Protractor No. 2: Vertical glazing, CIE Overcast Sky (BRE Copyright, reproduced with permission. Full set available from CRC, 151 Rosebery Avenue, London EC1R 4GB)

Figure 11.13 Sky Component Protractor No. 3: Horizontal glazing, Uniform Sky (BRE Copyright, reproduced with permission. Full set available from CRC, 151 Rosebery Avenue, London EC1R 4GB)

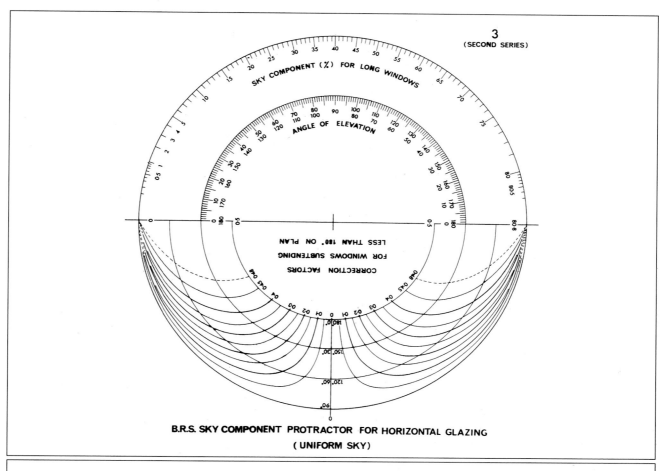

Figure 11.14 Sky Component Protractor No. 4: Horizontal glazing, CIE Overcast Sky (BRE Copyright, reproduced with permission. Full set available from CRC, 151 Rosebery Avenue, London EC1R 4GB)

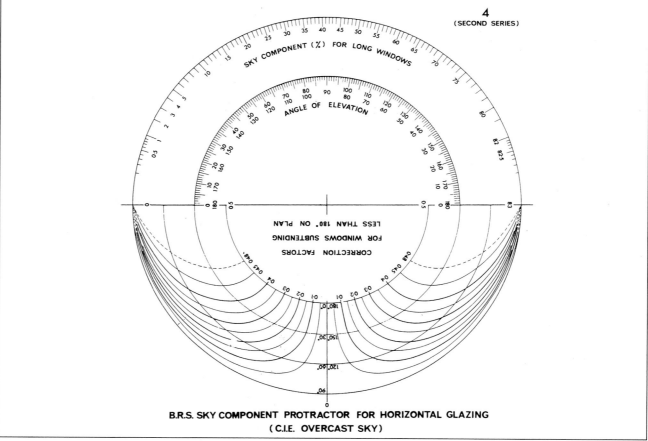

2 Calculate the area-weighted average reflectance of all the room surfaces, including the windows, using the following formula:

average reflectance = (floor area x floor reflectance +
wall area x wall reflectance +
ceiling area x ceiling reflectance +
window area x window reflectance)
÷ total room suface area

3 Locate the average reflectance value on Scale B and lay a straight-edge from this point across to the point already located on Scale A.

4 Where this line intersects Scale C read off the value, which gives the average IRC if there is no external obstruction.

5 If there is an obstruction locate its elevation angle from the centre of the window on Scale D.

6 Lay the straight-edge on this point on Scale D through the point already located on Scale C. The intersection with Scale E will give the corrected IRC taking account of the obstruction.

The nomogram evaluates the IRC assuming a ground reflectance of 0.1. For higher ground reflectances the IRC is increased, but since the ground-reflected light initially strikes the ceiling, the increment is dependent on the ceiling reflectance also. The correction graph can be used to evaluate a correction factor with which to multiply the IRC from the nomogram.

If nominal reflectances are used, e.g. from manufacturers' data, a further correction factor of 0.8 should be applied to the IRC to account for dirtying of surfaces.

The minimum IRC can be obtained by multiplying the average IRC by a factor (given in Table 11.6) that is dependent on the average room reflectance.

Table 11.6: Conversion factors to obtain minimum IRC.

Average reflectance	Conversion factor
0.3	0.54
0.4	0.67
0.5	0.78
0.6	0.85

Values of DF obtained from summing the SC, ERC and IRC calculated from the procedure above will be for normal single glazing. For different glazing systems the value should be multiplied by the ratio of light transmittance for the actual glazing used to that for single glazing. Transmittances for various glazing types are given in 12.3.

Allowance should also be made for dirtying of the glass surface, the DF being multiplied by the factors listed in Table 11.7.

Table 11.7: Conversion factors for location and maintenance.

Location + maintenance	Correction factor
Clean	0.9
Industrial	0.7
Very dirty	0.6

Figure 11.15
Nomogram for average internally reflected component for ground reflectance of 0.1 with correction factor for other ground reflectance values.

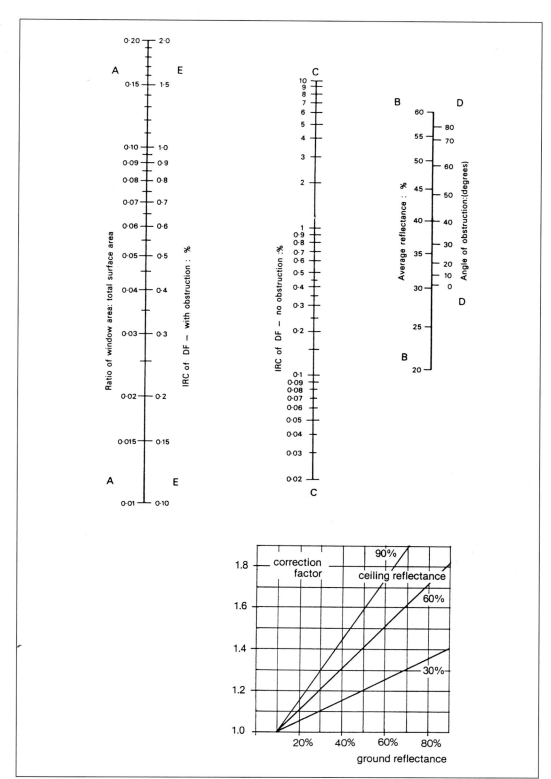

11.7 Sunpath diagrams

The cylindrical sunpath diagrams for latitudes 36°N, 40°N, 44°N, 48°N, 52°N and 56°N are shown in Figures 11.16 to 11.21. Their use for analysing the overshading by landscape and building obstructions has already been described in 4.4. Figure 11.22 is a shading mask. This can be photocopied onto acetate or tracing paper, and laid over the appropriate sunpath diagram with the arrow on the shading mask at the orientation of the window or the facade, on the azimuth scale of the sun path diagram (Figure 11.23). The droop lines of the shading mask are labelled with the shading angle. The shading angle is defined as the angle of elevation, in a plane at right angles to the facade, of the overhanging obstruction or ground-based obstruction such as building or wall. The droop lines represent the limit of the obstruction if the building edge runs parallel to the facade, such as the lower edge of an overhang or a boundary wall. The limits of azimuth angle of the obstruction in plan can also be represented on the diagram.

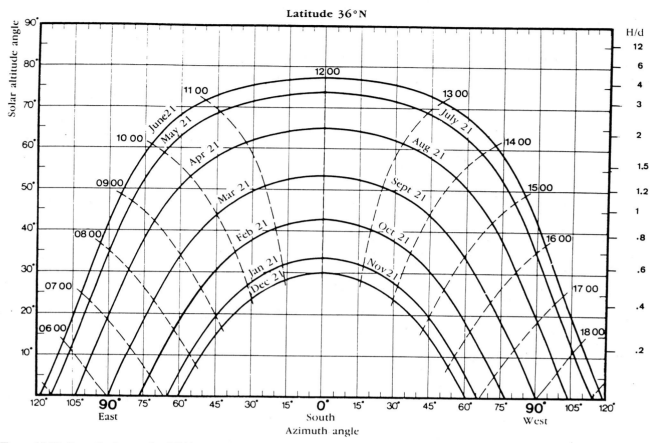

Figure 11.16 Sunpath diagram for 36° N

Figure 11.17 Sunpath diagram for 40° N

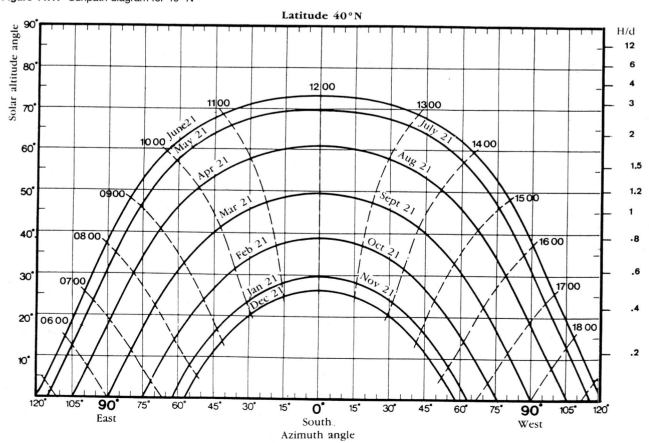

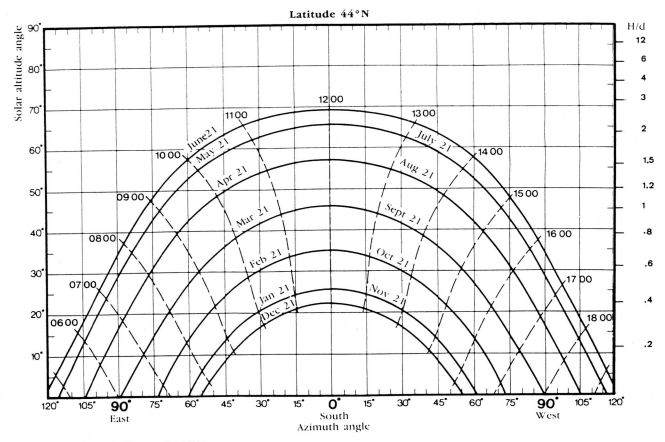

Figure 11.18 Sunpath diagram for 44° N

Figure 11.19 Sunpath diagram for 48° N

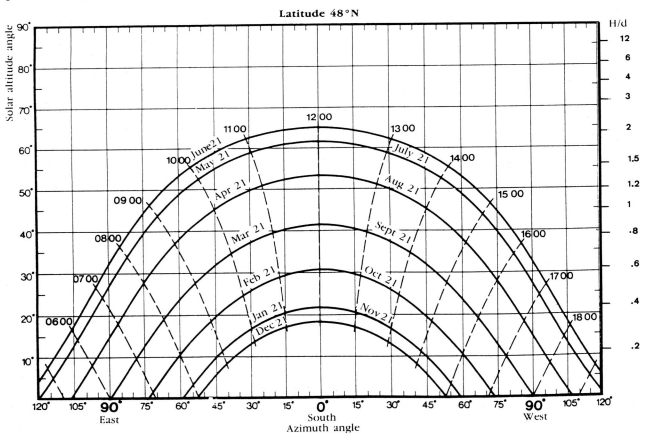

Figure 11.20
Sunpath diagram for
52° N

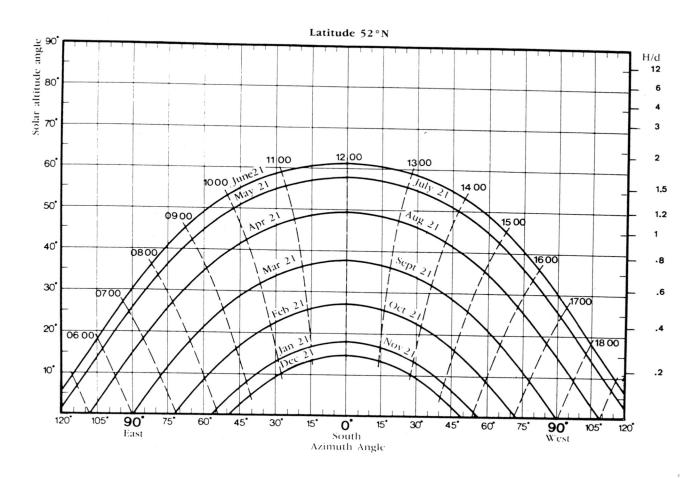

Figure 11.21
Sunpath diagram for
56° N

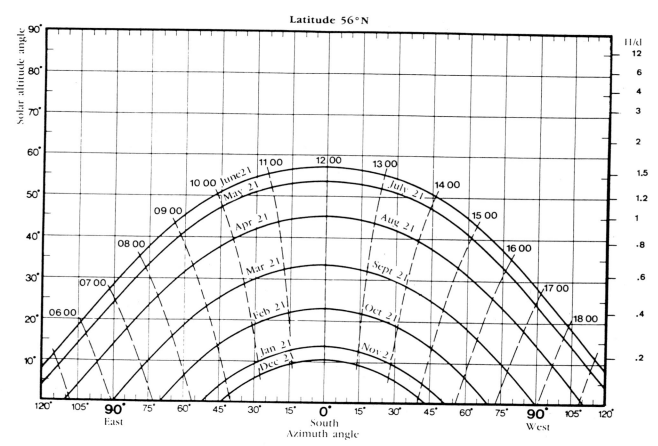

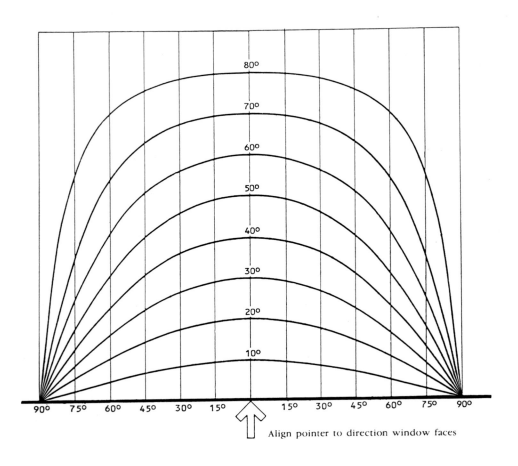

Figure 11.22 Shading mask for use with sunpath diagrams.

Align pointer to direction window faces

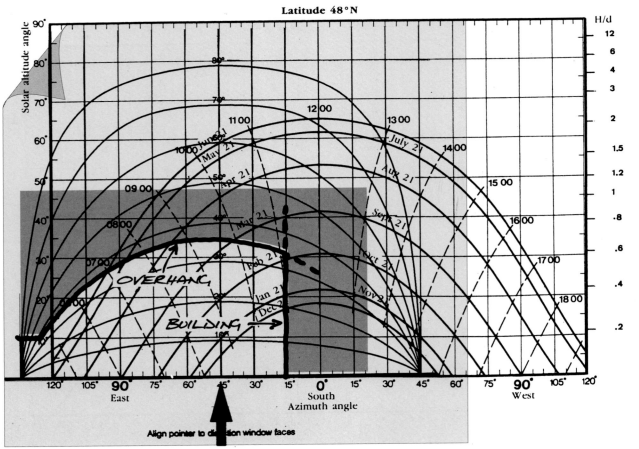

Figure 11.23 The shading mask has been positioned for a south-east-facing façade, i.e. at 45° from south (0°). The underneath edge of an overhang with a shadow angle of 35° has been sketched in, as well as a vertical line representing an obstruction at an angle of 15° from south. The sunpath diagram shows when the sun will be visible in the unshaded part of the sky.

References

1. N V Baker, A Fanchiotti and K Steemers (eds), *Daylighting in Architecture: A European Reference Book* (London: CEC DGII/James & James, 1993).

2. A McNicholl and J O Lewis (eds), *Daylighting in Buildings* (EC DG XVII Thermie, Energy Research Group, University College Dublin, 1994).

3. P Kenny and J O Lewis (eds), *Tools and Techniques for the Design and Evaluation of Energy Efficient Buildings*, EC DG XVII Thermie Action No. B184 (Energy Research Group, University College Dublin, 1995).

4. M E Aizlewood and P J Littlefair, "Daylight prediction methods: a survey of their use", *Conference Papers, CIBSE National Lighting Conference,* Bath, 1996, pp. 126–140.

5. B Paule and J-L Scartezzini, "Leso-DIAL: a new daylighting computer-based design tool", *Proceedings Right Light Four*, Copenhagen, 1997.

6. A Tsanggrassoulis, *PASSPORT-LIGHT*, National Observatory of Athens.

7. Windows and Daylight Group, *SUPERLITE 1.0: Program Description and Summary*, LBL Report DA-205 (California: Lawrence Berkeley Laboratory, July 1985).

8. Fraunhofer-Institut für Bauphysik, http://www.IBP.FhG.de/wt/Adeline/.

9. G J Ward, "RADIANCE lighting simulation and rendering system", *Proceedings SIGGRAPH '94, Computer Graphics Annual Conference Series*, July 1994.

10. M Fontoynont, *Proceedings ISES Conference*, Hamburg, 1987.

11. J Ashmore and P Richens, "Computer simulation in daylight design: A comparison", *Architectural Science Review,* vol 44, no 1, 2001.

12. E Tuchschmid and W Mathis W, "Etudes sous maquettes effectuées sous ciel artificiel", *Schweizerische Technische Zeitschrift*, No. 38/39, September 1966.

13. R G Hopkinson, P Petherbridge and J Longmore, *Daylighting* (London: Heinemann, 1966).

14. Tuchschmid and Mathis, "Etudes sous maquettes effectuées sous ciel artificiel".

15. M Schiller (ed.), *Simulating Daylight with Architectural Models* (Los Angeles: Daylighting Network of North America, University of Southern California, 1989).

16. M Szerman and H Künstlicher, "Quantifizierbare Tageslichtplanung im Entwurfsstadium", IBP Mitteilung, 174, 16, 1989.

17. P R Tregenza, "Daylight measurement in models: new type of equipment", *Lighting Research and Technology*, vol. 21, no. 4, 1989, pp. 193–194.

18. P R Tregenza, "Guide to recommended practice of daylight measurement: general class stations", Supplement to *CIE Journal*, vol. 6, no. 2, 1987.

19. Tregenza, "Daylight measurement in models".

20. L Michel, C Roecker and J-L Scartezzini, "Performance of a new scanning sky simulator", *Lighting Research and Technology*, vol 27, no. 4.

21. P R Tregenza, "Subdivision of the sky hemisphere for luminance measurements", *Lighting Research and Technology*, vol. 19, 1987.

22. J-L Scartezzini, L Michel, C Roecker and R Rhyner, *Laboratoire de Lumière Naturelle*, Projet OFEN (Lausanne: EPFL, 1994).

23. L Michel, R Compagnon and J-L Scartezzini, "Diagnostic in-situ du confort visuel au moyen d'un système vidéo", CUEPE/LESO-PB, *Proc. du 7 Energieforschung im Hochbau Status-Seminar*, September 1992, ETZ-Zurich.

24. Commission Internationale de l'Eclairage, *Daylight*, Publication CIE No. 16 (E-3.2) (Paris: CIE, 1970).

25. P Moon and D Spencer, "Illumination from a nonuniform sky", *Illuminating Engineering*, vol. 37, 1942, pp. 707–726.

26. Commission Internationale de l'Eclairage, *Standardisation of Luminance Distribution on Clear Skies*, Publication no. CIE 22 (TC 4.2) (Paris: CIE, 1973).

27. K Matsuura, *Luminance Distributions of Various Reference Skies*, CIE Technical Report of TC 3.09 (Paris: CIE, 1987).

28. B Molineaux, P Ineichen and O Guisan, *Mesures d'éclairage à Genève* (GAP/CUEPE, Université de Genève, 1993).

12 Design data

12.1 Building codes, daylight standards

The data given in table 12.1 is the result of a survey carried out in 1997 as part of the Daylight Europe project.

Table 12.1: Building codes – daylight standards

Country	Building codes	Standards, norms
Belgium		NBN L 13-002: *Daylighting of buildings. Determination of natural illuminance under overcast sky conditions.* (Consisting mainly of a translation of the CIE publication: No 16, 1970)
Denmark	Danish Building Code 1995	Ministry of Labor, Order No 1163, 1992 (Window area $\cong$ 10% of floor area by sidelight, or 7% by toplight
France	Code du Travail (Work Code) Code de la Construction et de l'Habitation (Building and Housing Code) Code de l'Urbanisme (City Planning Code)	NF X 35-103: *Ergonomical principles applicable to the lighting of workplaces for visual comfort* NF X 35-121: *Work on visual display unit and keyboard – Fittings of the work premises and the workplace*
Germany	Landesbauordnungen (State Building Codes); each German state has one of its own Verordnung Über Arbeitstätten / Arbeitstättenrichtlinien (Workplace Regulations), Nation-wide Tageslicht in Innenräumen, DIN 5034, Deutsches Institut für Normung e.V. (Daylighting in Interiors, DIN 5034, German Industrial Standards) – General Standards. Considering daylighting, the building codes and workplace regulations refer to this Standard. Consists altogether of 6 parts nowadays.	General requirements: Definition of terms / psychological importance of windows / Indoor illumination with daylight / Direct sunlight and sun protection Principles: Definition of terms / Astronomical basics / Radiation basics / Calculations: Daylight factor / Sunshine duration / Occupation time Simplified determination of minimum window sizes for dwellings Measurements: Definition of quantities to measure / Measurement equipment / Preparation of measurements / Simplified determination of suitable dimensions for roof lights
Greece	Greek Building Regulations (Ministry Decision 3046/304/30,1/3.2.1989, Government Official Gazette 59'	All spaces for primary use have to be provided with an aperture to the exterior of a min. size of 10% of the floor area The main staircase of building has to have natural lighting
Ireland	1991 Building Regulations, T. G. Part I: Conservation of fuel & energy (government regulations)	*Energy Efficient Lighting*, Eolas (The Irish Science and Technology Agency) BS.8206-part 2 code of practice for daylighting *Site Layout Planning for Daylight and Sunlight* – BRE PJ Littlefair
Italy	Ministry Decree 05/07/75 for dwellings: glazed area min. 12.5% of floor area; mean value of df $\geq$ 2% Ministry Decree 18/12/75 for schools: schoolrooms and laboratories df = 3%; gymnasia df = 2%; corridors, stairs and bathrooms df = 1%	Memorandum No 71911/10.0.296 , guidelines for VDU, Feb 1991 Legislative Decree No 626, Sept 1994

Norway	Tekniske forskirfter til plan og bygningsloven, 1997 Access to daylight mandatory for dwellings and workplaces Min. 1% df halfway into room, 1m from side wall, 0.8m above floor Glazed area min. 10%	Löfberg: Räkna med dagsljus Swedish Standard SS 914201: Dagsljus – Förenklad metod för kontroll av erfordelig fönsterglasarea
Portugal	General Regulation on Urban Buildings, 1951 (Decree-Law No 38382)	*Technical Norms for the Design of Residential Buildings*, Lisbon, LNEC, 1994 (Section 6.3.5: Visual Comfort)
Spain	Daylight building code rules in Spain are quite old. Usually they are only related to fenestration surface as a minimum (12.5% of the floor surface for living spaces)	
Sweden	There are no strict building codes, they just say 'adequate daylight' for dwellings (code issued by Boverket) The code recommends a glass area of min. 10% of floor area. With more than 20° horizon screening the glass area is recommended to be increased according to SS914201	SS914201: *Dagsljus – Förenklad metod för kontroll av erfordelig fönsterglasarea* Löfberg: *Räkna med dagsljus* Recommendation: The earlier building code stated 'at least 1% daylight 1m from side wall, at table high (0.8m above floor), halfway from windows to inner wall'. This is still regarded as reasonable goal even it is no longer clearly expressed in the new code. The value is for dwellings, but has also been used for offices, schools, nurseries etc.
Switzerland	Federal law for working spaces (10% minimal glazed ratio, 1/16 external view) Cantonal and district specific laws	ASE 8911.1989: *Indoor illumination by the way of daylighting* SIA 3804: *Electrical installation within buildings*
United Kingdom	For ventilation purposes the opening area of windows should be 1/20 the floor area	BS8206 Part 2: *Code of practice for daylighting* Site layout planning for daylight and sunlight

12.2 Daylight availability

The statistical description of illuminance from real skies throughout the year has already been introduced in 2.1.1. Sky illuminance on a horizontal plane is usually presented as a cumulative frequency curve of the percentage of the day (either sunrise to sunset or a specified working day) for which a given horizontal illuminance is exceeded, over the year.

The data presented here is an extract from the *European Daylighting Atlas*, produced within the framework of the DAYLIGHT II programme of the European Commision, completed in 1996. It is reproduced by kind permission of the co-ordinator, Prof D N Asimakopoulos, Technical University of Athens.

Data are presented for five European zones, which are indicated in Figure 12.1. The cumulative frequency curves are in Figures 12.2 – 12.6.

Orientation factors are also given for these zones. The use of these factors, which take account of the assymetry of the sky with respect to orientation, is described in

3.2.1 but is repeated briefly here.

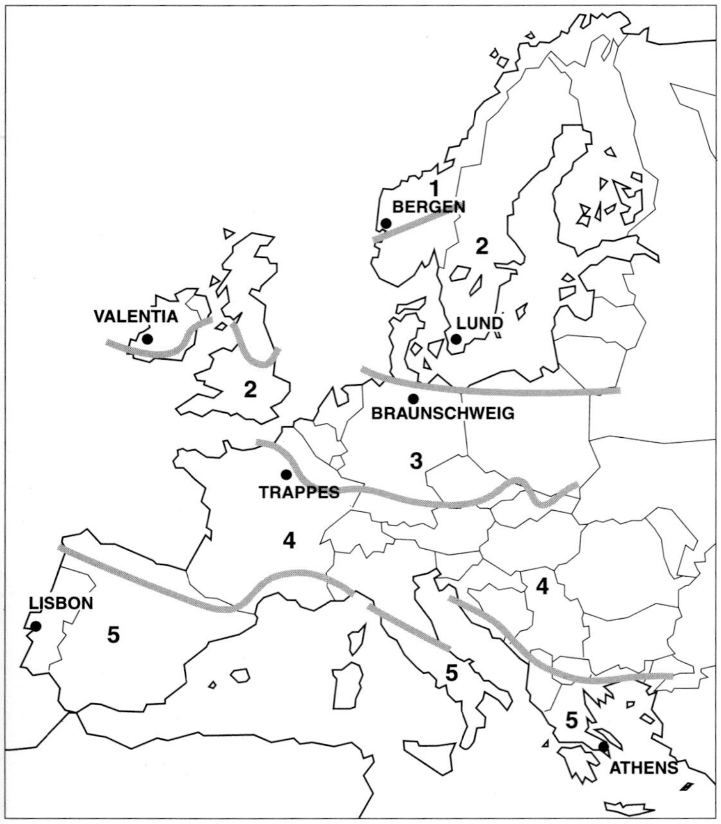

In establishing a target daylight factor (DF_t) to meet an illuminance criterion E_c for a given fraction $S\%$ of the year:

Step 1   From the daylight availability curve determine the outdoor illuminance E_S exceeded for $S\%$.

Step 2 $DF_t = 100 \times E_c / (E_s \times f_o)$,

where f_o is the orientation factor for the appropriate orientation of the major glazing area of the room, or an interpolated value. Note that values of f_o greater than 1 have the effect of increasing the external illumination, responding to the brighter illuminance of that part of the sky.

Note also that orientation factors are dependent on the occupancy period and are given for sunrise to sunset, 0900 – 1700 hrs and 0800 – 1400, the latter two corresponding to office and school hours respectively.

Example

What is the minimum DF that will achieve 65% sufficiency for an internal illuminance datum of 300 lux, in a room located in Daylight Zone 4? The room is occupied from 0900 to 1700 and has all its glazing orientated: (a) south and (b) east.

From Figure 12.5 $E_{65} = 17.5$ klux
therefore for (a) DF = 100 × 300 / (17 500 × 1.68)
                              = 1.0%

and for (b)   DF = 100 × 300 / (17 500 × 1.26)
                              = 1.4%

Figure 12.1 Daylight zones of Europe (*European Daylighting Atlas*. EU Contract JOU2 - CT92 - 0144 Coordinated by D. N. Asimakopolis, National Observatory of Athens, 1996).

**Typical hours
of use, all year**

Orientation Factors				
	N	**E**	**S**	**W**
Day	1.14	1.39	1.45	1.25
9-17	1.04	1.27	1.60	1.15
8-14	1.05	1.50	1.61	1.05

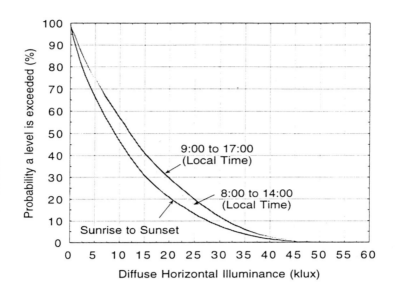

Figure 12.2 Frequency curves for diffuse horizontal illuminance in Bergen, Norway, Zone 1 (*European Daylighting Atlas.* EU Contract JOU2 - CT92 - 0144 Coordinated by D. N. Asimakopolis, National Observatory of Athens, 1996).

**Typical hours
of use, all year**

Orientation Factors				
	N	**E**	**S**	**W**
Day	1.11	1.30	1.47	1.32
9-17	1.01	1.19	1.66	1.23
8-14	1.02	1.38	1.63	1.05

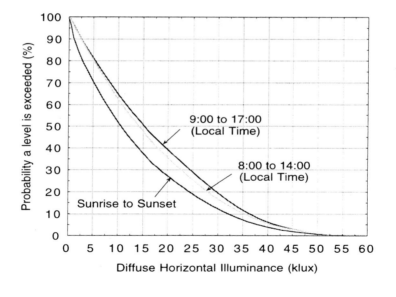

Figure 12.3 Frequency curves for diffuse horizontal illuminance in Lund, Sweden, Zone 2 (*European Daylighting Atlas*).

**Typical hours
of use, all year**

Orientation Factors				
	N	**E**	**S**	**W**
Day	1.06	1.23	1.40	1.25
9-17	0.96	1.14	1.56	1.14
8-14	0.97	1.34	1.52	0.99

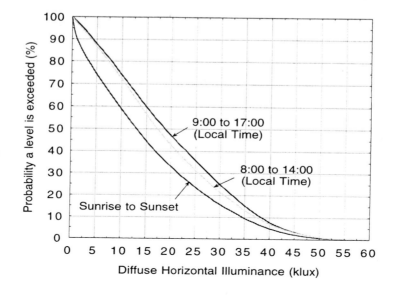

Figure 12.4 Frequency curves for diffuse horizontal illuminance in Braunschweig, Germany, Zone 3 (*European Daylighting Atlas*).

Typical hours of use, all year

Orientation Factors			
	N	**E**	**S**	**W**
Day	1.04	1.23	1.38	1.24
9-17	0.96	1.19	1.52	1.11
8-14	0.98	1.46	1.46	0.98

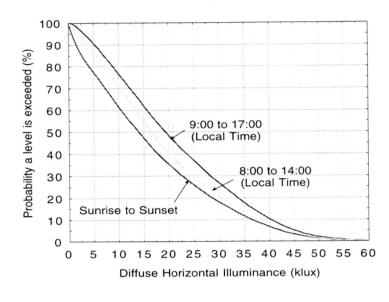

Figure 12.5 Frequency curves for diffuse horizontal illuminance in Trappes, France, Zone 4 (*European Daylighting Atlas*).

Typical hours of use, all year

Orientation Factors			
	N	**E**	**S**	**W**
Day	1.11	1.41	1.56	1.42
9-17	0.97	1.26	1.68	1.25
8-14	1.01	1.67	1.67	1.02

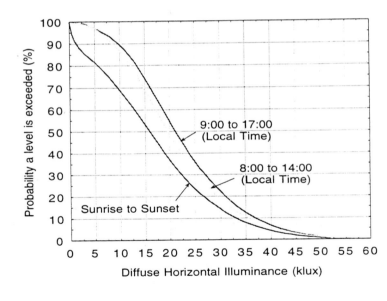

Figure 12.6 Frequency curves for diffuse horizontal illuminance in Athens, Greece, Zone 5 (*European Daylighting Atlas*).

12.3 Photometric data for materials

Most photometric data varies widely in its range of values even for the same nominal description. Thus it is not possible to give a precise reflectance for off-white paint or varnished softwood. This is equally true of glazing materials, where a description such as reflective blue glass would not define a unique set of data.

Thus these values are indicative only, where possible giving a range of typical values. For detailed design specification it is essential to consult manufacturers' product information. This is particularly true for glazing products, where a very large range of product types and variants exist.

Some of the photometric parameters have already been discussed in Chapter 6. However, for convenience, these and others relating to solar control glass are repeated here.

Reflectance: ratio of reflected to incident radiation

Diffuse reflectance: as above but for all directions of incident and reflected light

Transmittance: ratio of transmitted to incident radiation

Diffuse transmittance: as above but for all angles of incident and transmitted light

All above may carry the qualification visible or total.

Shading coefficient (short wave): the fraction of incident energy carried in the direct beam after passing through the glazing transmitted as short-wave radiation (visible and IR), as a proportion of the total energy transmitted by clear single glazing.

Shading coefficient (long wave): the fraction of incident energy that is transmitted into the room by long-wave radiation and convection from the inner surface of the glass as a proportion of the total energy transmitted by clear single glazing.

Total shading coefficient: the fraction of the total incident energy that is transmitted into the room, compared with clear single glazing; it is equal to the sum of the short- and long-wave coefficients.

Light/heat ratio: the ratio of the visible transmittance to the total shading coefficient.

Table 12.2: **Diffuse reflectance of building surfaces.**

Surface type	Description	Reflectance
Ceilings	White emulsion paint on plain plaster surface	0.8
	White emulsion paint on acoustic tile	0.7
	White emulsion paint on no-fines concrete	0.6
	White emulsion paint on wood-wool slab	0.5
Walls	White emulsion paint on plain plaster surface	0.8
	Tiles: white glazed	0.8
	Brick: white gault	0.7
	Plaster, pink	0.65
	White asbestos cement	0.4
	Brick: concrete, light grey	0.4
	Portland cement, smooth	0.4
	Stainless steel	0.35
	Brick, fletton	0.3
	Concrete: light grey	0.25
	Portland cement, rough (as board marked)	0.25
	Brick, London stock	0.25
	Timber panelling: light oak, mahogany, gaboon	0.25
	Timber panelling: teak, afromosia, medium oak	0.2
	Brick: concrete, dark grey	0.2
	Brick: blue engineering	0.15
	Chalkboard, painted black	0.05

Floors and furniture	Paper, white	0.8
	Cement: screed	0.45
	PVC tiles: cream	0.45
	Carpet: light grey, middle buff	0.45
	Timber: birch, beech, maple	0.35
	Timber: oak	0.25
	PVC tiles: brown and cream marbled	0.25
	Carpet: turquoise, sage green	0.25
	Timber: iroko, kerning, medium oak	0.2
	Tiles: cork, polished	0.2
	Quarry tiles: red, heather, brown	0.1
	Carpet: dark, 'low maintenance'	0.1
	PVC tiles: dark brown	0.1
	Timber: dark oak	0.1
Other	Asphalt	0.07
	Moist earth	0.07
	Salte (dark grey)	0.08
	Gravel	0.13
	Water	0.15
	Grandolite	0.17
	Bluestone, sandstone	0.18
	Macadam	0.18
	Vegetation (average)	0.25
	Cement	0.27
	Dark red glazed brick	0.3
	Green grass	0.33
	Dark buff brick	0.4
	Light buff brick	0.48
	Concrete	0.05–0.5
	Marble (white)	0.45
	Oak	0.15–0.05
	Old white paint	0.55
	New white paint	0.75
	Old snow	0.64
	New snow	0.74
Specular reflectance	Aluminium commercial grade (anodised and polished)	0.7
	Aluminium super-purity (anodised and polished)	0.8
	Surface aluminised glass or plastic	0.94
	Chromium (plate quality)	0.65
	Stainless steel (polished)	0.6
	Steel: white paint glossy (specular only)	0.05

Sources: CIBSE, 1994 *Code for InteriorLighting,* CIBSE, London.; Baker, N, Fanchiotti A and Steemers K; (eds), 1993 *Daylighting in Architecture,* James & James, London.
Cayless M.A and Marsden AM (eds); 1983 *Lamps and Lighting,* Edward Arnold, London.

Table 12.3: Typical light transmittance and reflectance for glazing.

Glass type		Light transmittance	Reflectance
Single pane			
Clear glass 6 mm		0.89	0.08
Clear glass 6 mm low-e		0.84	0.10
Clear acrylic 6 mm		0.92	0.08
Diffusing white acrylic 6 mm		0.17–0.72	–
Clear polycarbonate		0.83	–
Double pane with clear inner			
Clear glass 6 mm		0.76	0.10
Tinted 4–12 mm			
	Green	0.53–0.70	0.10
	Bronze	0.24–0.55	0.10
	Grey	0.17–0.49	0.07
Reflective 6–10 mm			
	Silver	0.09–0.26	0.38–0.16
	Bronze	0.09–0.22	0.19–0.17
	Blue	0.18–0.35	0.19–0.11
	Green	0.23–0.53	0.13–0.15
Double glazing for high thermal performance			
Aerogel and transparent insulation glazings			
	Monolithic aerogel	0.84–0.87	
	Polycarbonate capillary transparent insulation materials (TIM)	0.77–0.78	
	Polycarbonate honeycomb TIM	0.84	
Evacuated glazing			
	One low-e (0.2)	0.75	
	One low-e (0.1–0.2)	0.68	
Glazing with integrated blinds			
Lamellae open		0.5–0.08	
Lamellae closed		0.05–0.15	
Triple glazing			
Triple glazing		0.55	
Triple-glazed unit, 2 low-e (argon gas fill)		0.62–0.67	
Triple-glazed unit, 2 low-e (krypton gas fill)		0.63	

Source: abstracted from manufacturer's data.

Note 1: Values indicative only; do not use for design specification – refer to manufacturer's data.

Note 2: Values are for normal incidence. For diffuse transmittance x 0.91, for diffuse reflectance x 1.1.

Table 12.4: Typical shading performance for various glazing systems.

Glass type	Light transmittance	Shading coefficient (short wave)	Shading coefficient (long wave)	Total shading coefficient	Light transmittance total shading ratio
Single glazing 6 mm					
Clear glass	0.87	0.9	0.05	0.95	0.92
Body tinted					
Bronze	0.46	0.51	0.18	0.69	0.66
Grey	0.39	0.51	0.18	0.69	0.57
Green	0.66	0.53	0.17	0.70	0.94
Strongly reflecting	0.18	0.08	0.13	0.21	0.96
Wired cast	0.74	–	–	–	–
Double glazing					
6 mm clear and 6 mm:					
clear	0.76	0.70	0.12	0.82	0.92
low emissivity neutral	0.63	–	–	–	–
reflecting	0.26	0.29	0.09	0.38	0.68
strongly reflecting	0.15	0.13	0.07	0.20	0.75
High performance double glazing (6 mm + 6 mm)					
Reflective outer and clear float inner pane					
Neutral	0.52	0.39	0.11	0.50	1.04
Gold	0.49	0.27	0.08	0.35	1.40
Silver	0.42	0.26	0.07	0.33	1.27
Reflective outer and inner pane					
Green	0.35	0.15	0.14	0.29	1.20
Bronze	0.24	0.15	0.14	0.29	0.92

Source: abstracted from manufacturer's data.

Note: Light transmittance is for normal incidence. For diffuse transmittance x 0.91.

12.4 Light source and luminaires

The energy efficiency of an artificial lighting system is influenced by three key parameters:

- luminous efficacy of the source

- light output characteristics of the luminaire

- the geometry and reflectance of the room

Since the energy performance of the artificial lighting system will influence the value of daylight use it is useful for the designer to be aware of the characteristics of lighting hardware. Furthermore, where artificial lighting is to be used to supplement daylighting, compatibility of colour temperature is important. These characteristics will be found in the data below.

The dependence on the third parameter, room geometry and reflectance, is described by a utilisation factor. Typically this may vary from 0.2 to 0.8. Unfortunately this is not independent of the characteristics of the luminaires. This can be understood by considering the following.

A narrow-angle downlighter will deliver most of its output to the workplane below and thus will not be influenced by the reflectance of the walls or ceiling. In contrast, for a diffuse source emitting light in all direc-

tions, room reflectance will have a strong influence on illumination of the workplane since at least 50% of the light is reflected from room surfaces.

The illumination on the workplane can be calculated from

$$E = \frac{N.UF.M}{A}$$

where E is illuminance, N is the number of luminaires, UF is the utilisation factor, M is a maintenance factor ranging from 0.60 (dirty) to 0.83 (clean), and A is the area of the room. The formula does not include the light output ratio explicitly; this is included in the UF.

Manufacturers normally publish tables of utilisation factors for their luminaires for different room geometries and reflectances. Generalised values are published in the *CIBSE Code for Interior Lighting*. The topic of artificial lighting design in detail is outside the scope of this book.

Rather similar to the situation with glazing products, commercially available lighting products have a wide range of characteristics. The values listed here have been abstracted from trade literature. They are indicative only and should not be used for detailed specification. Manufacturers' data must be consulted.

Table 12.5: Luminous efficiency and colour rendering of principal lamp types.

Lamp type	Luminous efficiency (lm/W)	Colour temperature (K)	Colour rendering group
Filament lamps			
GLS and reflector	12	2700	1A
Tungsten-halogen	15–25	3000	1A
Fluorescent lamps			
Tubular warm	60–90	<3300	1A, 1B, 2, 3
Tubular Intermediate	60–90	3300–5300	1A, 1B, 2, 3
Tubular cold	60–90	>5300	1A, 1B, 2
Compact	45–90	2700–6500	1B
Induction lamps	70–90	3000–4000	1B
Low-pressure sodium lamps	100–185	–	–
High-pressure sodium lamps	60–100	2000–3000	1B–4
High-pressure mercury lamps	40–60	3300–4000	3
Metal halide lamps	60–80	3000–6000	1A–2

Source: CIBSE 1994, *Code for Interior Lighting*, CIBSE, London.

Table 12.6: Correlated colour temperature classes and colour rendering groups.

	Correlated colour temperature (CCT)	CCT class
	< 3300 K	Warm
	3300–5300 K	Intermediate
	>5300	Cold

Colour rendering groups	CIE general colour rendering index (Ra)	Typical application
1A	$R_a \geq 90$	Wherever accurate colour matching is required, e.g. colour printing inspection
1B	$90 \geq R_a \geq 80$	Wherever accurate colour judgements are necessary or good colour rendering is required for reasons of appearance, e.g. shops and other commercial premises
2	$80 \geq R_a \geq 60$	Wherever moderate colour rendering is required
3	$60 \geq R_a \geq 40$	Wherever colour rendering is of little significance but marked distortion of colour is unacceptable
4	$40 \geq R_a \geq 20$	Wherever colour rendering is of no importance at all and marked distortion of colour is acceptable

Source: CIBSE 1994, *Code for Interior Lighting*, CIBSE, London.

Table 12.7: Light output ratios of luminaires.

Luminaire type	ULOR	DLOR	LOR	
Surface-mounted fluorescent luminaires				
Trough reflector	0.05	0.70	075	
Rectangular reflector with opal diffuser	0.39	0.37	0.76	
Rectangular reflector with prismatic controller	0.39	0.48	0.87	
Polygonal luminaire body with white transverse louvre controller, 1 lamp	0	0.62	0.62	
Polygonal luminaire body with prismatic controller				
1 lamp	0.20	0.55	0.75	
2 lamps	0.10	0.54	0.64	
Mirror controlled with white transverse lamellae, 2 lamps	0	0.70	0.70	
Recessed fluorescent luminaires, 2 lamps				
Prismatic controllers	0	0.50–0.63	0.50–0.63	
Mirror controllers (low brightness)	0	0.61–0.70	0.61–0.70	
Mesh louvre	0	0.49–0.53	0.49–0.53	
Industrial discharge suspended luminaires				
With HPL lamp	0	074	074	
With HPL lamp and louvres	0	0.50	0.50	
With HPI or SON lamp	0	0.80	0.80	
With HPI lamp and louvres	0	0.62	0.86	

ULOR: upwards light output ratio

DLOR: downwards light output ratio

LOR: light output ratio

HPL: high-pressure mercury

HPI: metal halide

SON: high-pressure sodium

13 Post occupancy evaluation

13.1 Types of POE study

Post occupancy evaluation (POE) studies are carried out in buildings in order to systematically assess their performance, once they have been occupied and used. The aim is to determine whether the building meets the level of expectation that was envisaged in the conceptual stages of the design, in terms of both the human occupants and the building services.

The term POE has been used for studies that employ a wide variety of methods for collecting information on the use of a building by its occupants as well as on its environmental performance. Depending on these different methodologies, we can categorise POE studies as follows:

- objective observations of the physical environment

- objective observations of the occupant behaviour.

- subjective reporting by the occupant

13.1.1 Objective observations

In this case observations are made by the researcher. Data are derived by monitoring environmental parameters, often using instrumentation. An example would be the investigation of the use of artificial lighting in the building. The research team would monitor the hours when the system is switched on, as well as the use of the controls and possibly the prevailing daylight conditions. The data are then analysed and, based on this, the team will judge the success of the system and thus any possible need for its improvement.

The researcher may also make observations of occupant behaviour. This may be recorded descriptively, or may be logged quantitatively e.g. the number of times an occupant opens a window. Some of the latter type of observations may be carried out by technical instrumentation, rather than visual observation.

13.1.2 Subjective responses

One of the major concerns in the design of energy-efficient buildings is the judgement of the occupants on whether their environments are comfortable and acceptable for the activities they pursue. In this category of POE studies, the occupants of the building play the major role in providing the researchers with the informa-tion they are interested in. Their verbal responses are assessed and can indicate positive or negative aspects of the environmental performance of the building (e.g. day-lighting). In order to overcome the subjectiveness of this type of study and to obtain statistically significant results, the researchers include a large number of occupants or a constant group of people who will respond to the questions asked under different seasons throughout the year.[1] The methods used to collect the subjective responses of the occupants are:

- A questionnaire. Suggestions for typical questions in lighting questionnaires are discussed later.

- Through interviews with the occupants. These complement the questionnaire results, by allowing the occupants to elaborate on the answers they provided.

This introduces the issue of adaptation in the working environment. There is evidence, from research on environmental stress and adaptive behaviour, that people will engage in certain actions to achieve their environmental needs. In other words, a working area may be unsatisfactory, but often building occupants attempt to modify their environments in order to achieve more comfortable conditions. Research in thermal comfort theory indicates that an environment that may fail to satisfy conventional objective criteria may in practice prove satisfactory to the occupant because it provides the opportunity for adaptive behaviour, even if this action is not always taken.[2]

Both types of POE study are supported by data collected by the researcher on the building itself, such as plans and specifications. It may seem from the presentation above that each type of POE is very different from the other, but it is often the case that they are combined. For example, we mentioned above a study on artificial lighting use. If one wished to be thorough, one would need to include all three types of study. The environmental monitoring would provide the necessary quantitative data, a questionnaire would provide information on what the occupant perceived as a problem or as a positive aspect in his/her working environment, and finally the behavioural observations would show the researcher how the occupant actually uses the space.

13.1.3 Deliverables

By analysing the above data and by comparing the environmental monitoring results with the responses to the questionnaires and interviews, the researcher can identify issues which are of interest to the design team, management and those involved with the building's maintenance and, last but not least, the occupants of the space. Here is a list of the kind of interesting feedback that can be obtained from a lighting POE survey:

- Corrective actions that need to be taken. An example would be a case where the occupants complained of glare from the window, and sunlight penetration that interfered with their working activities. A proposition would be to rearrange the workplace layout or to include movable shading devices (e.g. blinds or curtains) on the windows in order to regulate the effect of the sky and the sun to improve the visual comfort of the occupants.

- Design features to be avoided. It is often the case that VDU needs prevent the location of large window areas nearby, due to glare and irritating reflections on the screen. The most frequent design solution is to isolate such users from the outside world, either by blocking the windows completely, or by allowing very small openings. However, questionnaire findings indicate the preference of users to have an outdoor view, and generally to be close to a window. A solution would be to regulate sky glare by appropriate window design and to always position VDU screens perpendicular to the external openings of the building.

- Achievement of targets. The researcher can establish whether design goals, energy targets (from environmental monitoring), standards and guidelines are actually achieved.

- Design aspects for long-term research and investigation. An example would be to check the energy performance of a specific lighting controls system and how successfully artificial lighting supplements daylight. This is often an initiator for more specific studies performed in buildings.

13.1.4 Interactive parametric studies

Apart from a general POE, which will gather information on user satisfaction in the work area and the environmental performance of the building, a second level of POE work involves parametric studies on specific aspects of the building that may have been found to function unsatisfactorily. In the former the research team did not introduce any changes to the environment. In interactive monitoring the researcher performs controlled experiments or intervention studies. In other words, the team identifies which aspect of the design of the building it wants to test and, by varying only that parameter, views the effects these changes have on the user of the space. If for example the researchers wish to test the response of people to peripheral lighting, they would choose a test site where they could manipulate the lighting arrangements and select a group of occupants who would assess different lighting arrangements.

This type of more specific POE work was undertaken for the Daylight Europe project. Staffan Hygge and Hans Allan Löfberg from the Kungl Tekniska Högskolan, Sweden, performed a number of POEs in eight buildings. Their findings are illustrated in the Case Study book.[3] One of these buildings is the LNEC building in Lisbon, Portugal, whose important feature is the external awnings. Different uses of these awnings were monitored and evaluated by a questionnaire issued twice and distributed to 80 people.

13.2 POEs for daylight

When the focus is daylighting, the aim is to evaluate the lighting quality of a building from a technical point of view and from the user's subjective point of view. The technical evaluations should concentrate on the energy aspects and function of the technical systems, such as controls, lighting, sunshading, etc. The user evaluations are derived from the questionnaires and/or interview. However, although primarily concerned with daylight, in order to have a more holistic approach towards the effect of daylighting in the building, the content of the questionnaire should consider the physical environment as a whole (also including noise and thermal conditions). Thus the user has the opportunity to rate the importance of lighting quality against other qualities of the work environment and work conditions.

Research recently carried out at the Martin Centre, University of Cambridge, following the methodology described above has produced interesting results regarding user preferences in library design.[4] The aim of the study was to compare findings from user surveys and monitoring data in order to illustrate positive and negative qualitative aspects of daylit interiors of Cambridge libraries, and thus produce guidelines for future use by professionals. The researcher selected a standard group of students for each library and asked them to fill in the same questionnaire under three different skies (summer sunny, winter sunny and overcast). At the same time she made objective environmental measurements. After gathering this information over a period of one year, the data were assessed using standard statistical tests.

The main result of this research was that daylighting quality, as perceived by the subjects, is affected by both quantifiable and non-quantifiable parameters. It was shown that subjects judge their daylighting environment depending on illuminance levels on the horizontal and vertical plane, while ratios (i.e., vertical-to-horizontal illuminance ratio) and the daylight glare index were less successful in predicting subjects' response. On the other hand the study identified a series of architectural parameters (view out, shape of room, surface reflectances, non-uniformity of luminances) and other non-technical parameters such as personal expectations and degree of adaptive opportunity (described below), which showed good correlation with the subjects' responses.

It was found that subjects not only enjoy having a view out of their working area, but are willing to accept high levels of glare in order to enjoy this privilege.

Regarding optimum reflectances, the study indicated that subjects preferred middle-range reflectances, while at the same time avoiding a bland uniform field of view. Variety is highly appreciated. Furthermore the shape of the room is also significant in how subjects judge their daylight environment; a combination of low reflectances and small spaces creates a feeling of darkness and gloominess.

A new method of presenting luminance variation in the 360° field of view demonstrated that subjects preferred non-uniformity in their surroundings but not so much between the average of their surroundings and their task. This variation in the surrounding field of view could be achieved not only by different wall reflectances, but also by room elements such as a bookstack or a window, thus creating visual interest as well as luminance variation.

Another significant result was that if subjects see a bright daylit scene through a window, they will delay switching on the lights, thus resulting in saving energy.

Finally satisfaction was shown to be strongly connected with availability of individual choice and control, referred to as adaptive opportunity. It is important to allow for user-friendly controls, while at the same time providing a variety of seating areas (in the case of libraries). The user will subsequently have the option to sit closer to the window and counteract possible visual problems by drawing the blinds, or sit farther away from the window wall and still enjoy a view out. Flexibility, options and interaction are the key words in a successfully daylit library interior.

Such research is beneficial, and should be extended in other type of buildings, since it provides the designers with a series of recommendations that will increase their knowledge and awareness on topics that often seem too 'scientific'. The ultimate goal should be to refine standards and recommendations with the results of POE research, so that during the design process day-lighting quality is not left solely to intuition and chance.

13.2.1 Questionaire

In general POEs, it is not important to be able to identify the person who is filling in the questionnaire each time. The researcher will need to be able to identify workspaces to facilitate the analysis of data on a locational basis. There are no general rules as to the exact number of people involved in the questionnaires. Generally the more participants the better, in order to avoid errors from subject variance. Furthermore it is useful to include an equal number of female and male occupants, if that is possible, and to divide them into categories depending on their eyesight, workspace and position in the company. There is no specific guidance for the precise content of a questionnaire, since each lighting design and each building in general is unique, and thus the questions may be different in each case. In order, though, to have a general idea of how such questionnaires are presented, a series of points are noted below.

The first part of the questionnaire includes general questions regarding:

- date and time of questionnaire

- age, sex, profession of occupant

- duration of occupant's work in the building and at his/her specific workspace

- type of work they are involved in

- type of workspace

These can be presented with a series of suggested responses (check lists), e.g.:

What type of workspace do you currently occupy? (check one)

[...]   An enclosed, private office

[...]   An enclosed office shared with others

[...]   A workstation with screens or other type of lighter partition

[...]   An office without partitions

[...]   Other (please specify):

The main body of the questionnaire may include one or several of the points noted below. For each point, different questions are indicated and a suggested type of response is noted, e.g. Yes/No. Although sample questions are given, these are as examples only in type and structure and are not meant to be definitive.

Relationship to windows

- Existence of window or skylight, or view of those from the workspace (Yes/No)

Sunlight discomfort

- Whether there is sunlight ever on the worktop (Yes/No)

- Whether they like the presence of sunlight in their field of view (Yes/No)

- How often they can control possible sunlight penetration with drapes or blinds (Yes/No)

- Whether they find it too bright to work.

Answers are generally given on a rating scale which varies from a multi-point scale to a simple line on which they tick, e.g.:

When the sunlight is bright, how much does it interfere with your work?
(circle one)

0 = Not at all

1 = A little

2 = Moderately

3 = Very much

or

(please tick)

Very much ___ : ___ : ___ : ___ : ___ Not at all

Daylight

- Questions on the presence of daylight in their workspace (Yes/No)

- Questions on how much they enjoy daylight in their workspace (Answer on a scale, as above)

- Questions on its adequacy (functionality) (Answer on a scale)

- Question on how often electric lighting from the ceiling or from a desktop light is used, in addition to daylight (Answer on a scale)

Electric light usage (controls)

- Questions on whether they can switch on and off ceiling lights and desk lights (Yes/No)

- Questions on whether they are aware of automatic controls (Yes/No)

- Questions on whether they are satisfied with their function (Answer on a scale)

- Questions on how often they ignore the sensors and switch the lights on or off (Answer on a scale)

Glare

Questions on this topic can either be incorporated in each of the above sections (e.g. sunlight or computer use (below)) or be a separate part of the questionnaire. Furthermore a small explanation of the term 'glare' may be needed in the beginning.

- Questions on how often they experience glare (Answer on a scale)

- Check list of possible sources of glare could be included.

- Questions on how much glare interferes with their work (Answer on a scale)

Computer use

- Whether they use one or not (Yes/No)

- Work position (check list)

- How much time they spend per day on the computer (check list)

- Whether they experience bothersome reflections on the screen (check list)

- What is the cause of the reflections (check list)

- How much they bother him/her (Answer on a scale)

Finally, if one is looking into adaptive behaviour, the questionnaire should include questions on how they cope with individual problems (e.g. computer screen reflections). This is referred to as the 'coping' list, which describes behaviours aimed at improving the environment, the subject's own behaviour (change of position etc), or how one thinks or feels about a situation. These can be presented as open-ended questions or with a series of suggested responses (check lists). An example from an American survey is noted below:[1]

When the sunlight on your workspace feels too warm, how do you respond?
(check all that apply)

[...]   Close the drapes / blinds

[...]   Open a door / window to the outdoors

[...]   Open an interior door / window

[...]   Drink something cool

[...]   Adjust the thermostat

[...]   Go outdoors for a while

[...]   Move to a more comfortable location in my workspace

[...]   Work elsewhere in the building

[...]   Try to ignore any discomfort and concentrate harder on my work

[...]   Just put up with it – there is nothing I can do

[...]   Other

References

1.   J H Heerwagen and J Loveland, *Energy Edge, Post Occupancy Evaluation Project, Workspace Satisfaction Survey*, Final report WA 98195 (Seattle: University of Washington, October 1991).

2.   N V Baker and M Standeven, "Thermal comfort in free running buildings", *Energy and Buildings*, no. 23, 1996.

3.   *Daylight Performance of Buildings* Ed Marc Fontoynont, James & James (Science Publishers) Ltd, London, 1999.

4.   K Parpairi, "Daylighting in architecture: quality and user preferences", Doctoral thesis, May 1999.

Annotated bibliography

American Architectural Manufacturers Association, AAMA (Ed.), *Skylight Handbook Design Guidelines*, AAMA, Illinois, USA, 1987.

Includes data on maximising skylight, energy and daylighting benefits in commercial buildings. Based on Lawrence Berkeley Laboratories research.

Ander, G.D., *Daylighting Performance and Design*, Southern California Edison, California, 1986.

Addresses pragmatic issues of daylighting with a number of case studies and appendices. It concludes with an annotated American bibliography.

Argiriou, A.A., Asimakopoulos, D.N., Balaras, C.A. (Eds.), *Daylight Techniques in Buildings, Reference Handbook*, The European Commission DGXVII Altener Programme, Athens, 1997.

Baker, N.V., Fanchiotti, A., Steemers, K. (Eds.), *Daylighting in Architecture: A European Reference Book*, James & James Ltd for the Commission of the European Communities, London, 1993.

This major work draws on European research and expertise and provides the reader with a good overview of daylighting techniques and research. Some rather specialist chapters.

Bell, J.A.M., Burt, W., *Designing Buildings for Daylight*, Building Research Establishment, Watford, 1995.

A guide that includes illustrated examples of successful daylit buildings and explanations of key issues.

Boyce, P.R., *Human Factors in Lighting*, Applied Science Publishers Ltd, London, 1981.

Very good book on experiments in artificially lit environments, testing the subjective responses of people under different lighting arrangements.

Building Research Establishment (BRE), *Energy Conservation in Artificial Lighting*, BRE, Watford, UK, 1979.

Recommendations on optimising the use of lighting power. These apply both to existing buildings and to designs for new buildings.

Building Research Establishment (BRE), *Office Lighting for Good Visual Task Conditions*, BRE, Watford, UK, 1981.

Discusses issues such as veiling reflections and their quantification with the use of the contrast rendering factor.

Building Research Establishment (BRE), *Lighting Controls and Daylight Use*, BRE, Watford, UK, 1985.

Based on behavioural studies, the traditional manual switching arrangements are discussed, as well as a method of predicting energy savings from automatic lighting controls.

Building Research Establishment (BRE), *Estimating Daylight in Buildings*, Parts 1 and 2, BRE, Watford, UK, 1986.

Part 1 discusses the sky component and externally reflected component, while Part 2 discusses the internally reflected component. It also provides explanation on the use of the BRS protractors and Waldram Diagram.

BRECSU, *Energy Efficient Lighting in Buildings*, BRE for the Commission of The European Communities DG XVII, Watford, 1991.

Maxibrochure on the energy implications of lighting in buildings in Europe. It includes 13 brief case studies.

Brown, G.Z., Cartwright, V., *Sun Wind and Light: Architectural Design Strategies*, John Wiley and Sons, New York, 1985.

As the title suggests, it is an interesting book on architectural design strategies.

Building Research Station (BRS), "Estimating daylighting in buildings: Parts 1 and 2", *Building Research Station Digest*, 42:1-7, January 1977.

Outline of the daylight factor method of analysis as first presented in Hopkinson's *Daylighting*.

Button, D., Pye, B., (Eds.), *Glass in Buildings: A Guide to Modern Architecture Glass Performance*, Butterworth Architecture, Oxford, UK, 1993.

Very good quality photographs and an interesting presentation of properties of different glazing materials.

Chartered Institution of Building Services Engineers, *Applications Manual: Window Design*, CIBSE, London, 1987.

A concise guide on window design both for new and for existing buildings. It discusses both rooflight and window design, glare, view and privacy issues and in the second part provides the reader with design information such as the daylight factor measurement. Contains many quantitative procedures.

Chartered Institution of Building Services Engineers, *Lighting for Offices, Lighting Guide LG7: 1993*, CIBSE, London, 1993.

Includes seven sections: introduction and scope, office lighting design criteria, lighting systems, lighting equipment, recommendations for specific applications, lighting design and inspection, servicing and maintenance of lighting installations.

Chartered Institution of Building Services Engineers, *Code for Interior Lighting*, CIBSE, London, 1994.

One of the most important concise texts on artificial lighting design. It discusses its visual effects, gives tables of recommendations depending on the building use, and includes a part on lamps, luminaires and control mechanisms.

Chartered Institution of Building Services Engineers, *Lighting for Museums and Art Galleries*, CIBSE Publications, London, 1994.

Gives specialist advice on the design and provision of lighting systems for the display of works of art.

Crisp, V.H.C., Littlefair, P.J., Cooper, I., McKennan, G., *Daylighting as a Passive Solar Energy Option: An Assessment of its Potential in Non-Domestic Buildings*, Building Research Establishment, Watford, 1988.

Discusses how to exploit daylight in order to improve energy efficiency in non-domestic buildings.

Energy Research Group, School of Architecture, University College Dublin (Ed.), *Energy in Architecture: The European Passive Solar Handbook*, B.T. Batsford for the Commission of the European Communities, London, 1993.

A general book on environmental issues with a section in daylighting.

Egan, M.D., *Concepts in Architectural Lighting*, McGraw-Hill for College of Architecture, Clemson University, New York and London, 1983.

Includes discussion on daylighting, design and lighting models.

Evans, B. H., *Daylight in Architecture*, McGraw-Hill, New York, 1981.

A design-oriented book which provides the reader with the necessary concepts and model testings. It is a good place to start when one is entering the field of daylighting.

Falk, D.S., Brill, B.R., Stork, D.G., *Seeing the Light*, John Wiley & Sons, New York, 1986.

A comprehensive book on optics, colour and vision, with excellent illustrations.

Gardner, C., Hannaford, B., *Lighting Design, An Introductory Guide*, The Design Council, London, 1993.

Although it refers mostly to artificial lighting, it makes reference to daylighting and environmental issues. It also includes a number of case studies.

Groupement Européen des Producteurs de Verre Plat (G.E.P.V.P.), *Natural Lighting in Architecture, Windows: Factors in the Quality of Life*, G.E.P.V.P., Brussels, 1987.

A free brochure on the advantages of daylighting, availability of daylight and methods for predicting illuminance.

Hopkinson, R.G., *Architectural Physics: Lighting*, Her Majesty's Stationery Office, London, 1963.

A very interesting book on subjective reactions and subjective studies. It tries to relate the study of the individual to practical lighting technology.

Hopkinson, R. G. (Ed), *Hospital Lighting*, Heinemann, London, 1964.

A series of papers on hospital lighting design with interesting but somewhat outdated photographs.

Hopkinson, R. G., Petherbridge, P., Longmore, J., *Daylighting*, Heinemann, London, 1966.

Probably the most valuable resource for daylighting and design methods. A book which must be read by anyone interested in this area of research. Contains good bibliographical reference at the end of each chapter. Although it is based on NW European conditions it incudes a chapter on tropical regions.

Hopkinson, R.G., Collins, J.B., *The Ergonomics of Lighting*, MacDonald &Co Ltd, London, 1970.

Addresses all issues on visual comfort and perception.

Hunt, D.R.G., *Availability of Daylight*, Building Research Establishment, U.K., 1979.

Technical summary of data on the outdoor illuminances recorded at different times of the day and throughout the year. It is used for predicting likely hours of artificial lighting use.

Illuminating Engineering Society of North America, *Recommended Practice of Daylighting*, IES, New York, 1979.

A very good source of daylighting information with charts and tables on the procedure of the IES method.

Lam, W.M.C., *Perception and Lighting as Formgivers for Architecture*, McGraw-Hill, New York, 1977.

The text investigates the psychology of visual perception supported by a series of case studies covering the second half of the book.

Lam, W.M.C., *Sunlighting as Formgiver for Architecture*, Van Nostrand Reinhold Company, New York, 1986.

Sunlighting design strategies are covered and supported with case studies. It introduces theoretical and practical aspects of sunlighting and looks into ways of reducing energy use and cost. The book ends with how to use physical models.

Lechner, N., *Heating, Cooling, Lighting: Design Methods for Architects*, John Wiley & Sons, New York, 1991.

It is a well illustrated book, easy to comprehend and includes guidelines on all three areas as well as design tools, rules of thumb and examples.

Littlefair, P.J., *Daylight, Sunlight and Lighting Control*, Building Research Establishment, Watford, UK, 1980.

A teaching package with slides and drawings. Topics included are daylight, sunlight, innovative daylighting and lighting controls.

Littlefair, P.J., *Solar Dazzle Reflected from Sloping Glazed Facades*, Building Research Establishment, Watford, UK, 1987.

It presents a method used to calculate, still at the design stage, whether solar dazzle will result from a proposed building facade.

Littlefair, P.J., *Average Daylight Factor: A Simple Basis for Daylight Design*, Building Research Establishment, Watford, UK, 1988.

Quick and accurate ave D.F. formulas. By using them one can determine target glazing areas.

Littlefair, P.J., *Designing with Innovative Daylighting Systems*, Building Research Establishment, Watford, UK, 1996.

Mirrors, prismatic glazing, lightshelves and light pipes are discussed as systems which maximise daylight usage in buildings.

Littlefair, P.J., Lindsay, C.R.T., *Scale Models and Artificial Skies in Daylighting Studies*, Building Research Establishment, Watford, UK, 1990.

It provides guidelines on daylight measurement for building models, under different artificial skies. Includes a list of UK artificial skies.

Littlefair, P.J., *Site Layout and Planning for Daylight and Sunlight: A Guide to Good Practice*, Building Research Establishment, Watford, UK, 1991.

It provides new guidance in this field and should be read with the interior daylight recommendations in BS8206 and the CIBSE *Applications Manual: Window Design*.

Littlefair, P.J., *Site Layout for Sunlight and Solar Gain*, Building Research Establishment, Watford, UK, 1992.

Gives guidance on achieving good sunlight access for buildings and open spaces between them.

Littlefair, P.J., *Measuring Daylight*, Building Research Establishment, Watford, UK, 1993.

Based on measurements made at the BRE illustrating the changing nature of light from the sky, advice is given on how to measure daylight under real and artificial skies.

Littlefair, P.J., *Daylighting Design for display-screen equipment*, Building Research Establishment, Watford, UK, 1995.

Describes how to permit daylight in the space so that the interior benefits from it while screen reflections and glare are avoided.

Lynes, J.A., *Principles of Natural Lighting*, Elsevier Publishing Company Ltd, London, 1968.

Covers all the basic issues on daylighting.

McNicholl, A., Lewis, J.O., (Eds.), *Daylighting in Buildings*, University College Dublin – OPET for the European Commission, Directorate-General for Energy (DGXVII), Dublin, 1994.

A good overview to daylighting principles and design.

Moore, F., Anderson, G., *Concepts and Practice of Architectural Daylighting*, Van Nostrand Reinhold, New York, 1985.

A comprehensive text on the basics of daylighting. Good graphics make the whole theory very understandable. The content relates to North America. The text is supported by a glossary and 7 appendices.

Muneer, T., *Solar Radiation and Daylight Models for Energy Efficient Design of Buildings*, Butterworth-Heinemann, Oxford, 1997.

A recent publication on daylight modelling.

Olgyay, A., Olgyay, V., *Solar Control and Shading Devices*, Princeton University Press, Princeton, New Jersey, 1976.

A classic book on shading devices. It begins with a historical overview and ends by outlining a detailed analysis and design process.

Pritchard, D.C., *Lighting*, Longman Scientific & Technical, Essex, 1995.

It covers both interior and exterior lighting, contains sufficient data for most lighting design calculations, and incorporates new material on lighting different types of interiors. It also includes worked examples to support the key concepts.

Rasmussen, S.E., *Experiencing Architecture*, The M.I.T. Press, Cambridge, Mass., 1964.

A unique architectural discussion of the relationship between daylight and architecture; appreciated by architects and non-architects alike.

Rea, M.S. (Ed.), *Lighting Handbook, Reference and Application*, 8th Edition, Illuminating Engineering Society of North America (IESNA), New York, 1993.

A general handbook on natural and artificial lighting, covering a wide variety of topics.

Robbins, C.L., *Daylighting: Design and Application*, Van Nostrand Reinhold Company, New York, 1986.

It is a two-part handbook on the fundamentals of daylighting. The first part incorporates control devices and analysis methods used in daylighting. The second contains reference material and data.

Schiller, M. (Ed.), *Simulating Daylight with Architectural Models*, Daylighting Network of North America, Los Angeles, 1988.

It provides the reader with the fundamentals of scale model techniques for daylighting studies, and covers aspects of their construction, testing and evaluation of results.

Slater, A.I., *Lighting Controls: An Essential Element of Energy Efficient Lighting*, Building Research Establishment, Watford, UK, 1987.

Discusses control strategies based on practical experience gained in large-scale installations.

French publications:

Association Française de l'Eclairage (A.F.E.), *La Photométrie en Eclairage*, Société d'Editions LUX, Paris, 1983.

Italian publications:

Torricelli, M.C., Sala, M., Secchi, S., *La Luce del Giorno. Tecnologie e Strumenti per la Progettazione*, Alinea, Firenze, 1995.

Swedish publications:

Andersson, O. (Ed.), *Rum och Ljus (Room and Light)*, Arkus, Stockholm, 1988.

Boverket BFS, *Nybyggnadsregler (Rules for new buildings)*, Boverket BFS 1988:18, Allmänna Förlaget, Stockholm, 1988.

Löfberg, H.A., *Dagsljus Utomhus (Daylight outdoors)*, Information sheet B9:1976, Statens råd för byggnadsforskning, Stockholm, 1976.

Löfberg, H.A., *Räkna med Dagsljus (Count on daylight)*, Statens Institut för Byggnadsforskning, Gävle, 1987.

Swedish Standard, *Byggnadsutformning – Dagsljus – Förenklad Metod för Kontroll av erforderlig fönsterglasarea (Building design – Daylighting – Simplified method for checking required window glass area)*, SS 91 42 01, Stockholm, 1988.

Swiss publications (in French, German and Italian):

Projet DIANE – Eclairage Naturel, *Savoir Conduire la Lumière Naturelle en Architecture – Quelques Pistes de Réflexion*, Office Fédéral de l'Energie, Berne, 1995 (in French and German).

Projet DIANE – Eclairage Naturel, *Systèmes Techniques pour l'Utilisation Optimale de la Lumière du Jour*, Office Fédéral de l'Energie, Berne, 1995 (in French and German).

Projet DIANE – *Eclairage Naturel, Tageslichtnutzung in Gabäuden – Denkanstösse*, Office Fédéral de l'Energie, Bern, 1995 (in French and German).

Projet DIANE – Eclairage Naturel, *Systeme der Tageslichtnutzung – Beispiele, Messungen, Tendenzen*, Office Fédéral de l'Energie, Bern, 1995 (in French and German).

OFQC/Bfk, *Eléments d'éclairagisme / Grundlage der Beleuchtung / Principi di illuminotecnica*, EDMZ, Berne, 1993 (in French, German and Italian).

OFQC/Bfk, *Eclairage dans l'industrie / Zeitgemässe Beleuchtung von Industriebauten*, EDMZ, Berne, 1993 (in French and German).

OFQC/Bfk, *Eclairage des bureaux / Zeitgemässe Beleuchtung von Bürobauten / Illuminazione degli uffici*, EDMZ, Berne, 1994 (in French, German and Italian).

OFQC/Bfk, *Eclairage des surfaces de vente / Effiziente Beleuchtung von Verkaufsflächen*, EDMZ, Berne, 1994 (in French and German).

OFQC/Bfk, *Neuer Komfort mit Tageslicht / Lumière naturelle à bon escient*, EDMZ, Berne, 1995 (in French and German).

SLG-LiTG-LTAG-NSVV, *Handbuch für Beleuchtung*, Ecomed, Landsberg, 1992 (in German).

SUVA / CNA, *Le Travail à l'Ecran de Visualisation*, CNA, Lucerne, 1991.

Journals which feature articles about daylighting:

Architectural Lighting (American)

Architectural Science Review (Australian)

Building and Environment (British)

Building Services Engineering Research and Technology (British)

Energy and Buildings (Swiss, in English)

International Association for Energy-Efficient Lighting (IAEEL) (Free newsletter published in Sweden)

Journal of the Illuminating Engineering Society (American)

Lighting Design and Applications (American)

Lighting Research and Technology (British)

Glossary

Accommodation: The ability of the eye to bring into 'sharp focus' objects at varying distances from infinity down to the nearest point of distinct vision, called the 'near point'. Age has a profound effect on accommodation ability because the eye's lens gradually loses its elasticity. As a result the near point gradually recedes (average near point at different ages: 16 years→8 cm, 32 years→12 cm, 44 years→25 cm, 50 years→50 cm, 60 years→100 cm).

Adaptation: The process by which the state of the visual system is modified by previous and present exposure to stimuli that may have various luminances, spectral distributions and angular subtenses. Note: the terms 'light adaptation' and 'dark adaptation' are also used, the former when the luminances of the stimuli are > 1 cd/m^2, and the latter when the luminances are < 0.01 cd/m^2.

Adaptation level: The luminance that results in an object having the same apparent brightness as it would have in a uniform field of that luminance.

Altitude (used to describe an angle): The angle between the horizontal plane and a specific direction. It is common to use this term to describe directions pointing towards the sky hemisphere or the sun. This angle is sometimes also called 'height'.

Anidolic daylighting systems: 'Anidolic' is a synonym of 'non-imaging', constructed from two words of ancient Greek (*an* = 'without', *eidolon* = 'image'). Although many daylighting systems can claim to be inherently 'non-imaging' because they are intended to channel natural light into a room without taking into account image distortions that may occur, the qualifier 'anidolic' applies only to those systems that have been designed using the concepts and tools found within the 'non-imaging optics' theoretical framework.

Artificial sky: An enclosure that simulates the luminance distribution of a real sky or standard design sky for the purpose of testing physical daylighting models.

Atrium: An interior space enclosed partially or completely by walls of a building and covered with transparent or translucent material. The interior spaces of the parent building are separated from the atrium space by walls and glazing.

Azimuth: Direction relative to the north. Sometimes presented relative to south.

Borrowed light: Secondary light coming from an adjacent space.

Brightness or apparent brightness: Attribute of a visual sensation according to which an area appears to emit more or less light compared with the average field. It is the subjective response to luminance in the field of view dependent on the adaptation of the eye.

Brightness constancy: The condition achieved under adequate illuminance where the perceived 'lightness' of objects is relatively unchanged through fairly large changes of illuminance.

Candela (cd): The SI unit of luminous intensity, equal to 1 lumen per steradian.

Chroma (saturation): A measure of the degree of vividness of a colour from a pure hue, through varying tints of decreasing colour, to grey. In the Munsell system the index varies from 0 for neutral grey to 10 or over for strong colours.

CIE standard overcast sky: A completely overcast sky for which the ratio of its luminance (Lγ) at an angle of elevation (α) above the horizon to the luminance (L$_z$) at the zenith is given by the formula:

$$L_\alpha = L_z \frac{(1 + 2 \sin \alpha)}{3}$$

This results in the sky being three times brighter overhead than at the horizon.

Clear sky: Sky unobstructed by cloud.

Clerestory: An area of glazing, usually vertical, well above normal window height. May be present above normal windows, or separate.

Colour adaptation: The property of the three primary colour receptors to adapt independently to compensate for different compositions of white light.

Colour constancy: Psychological mechanism of colour adaptation which makes us see a colour as we think it ought to be rather than as it is. The result of colour adaptation and similar to brightness constancy.

Colour rendering index (CRI): A measure of the degree to which the colours of surfaces illuminated by a given light source conform to those of the same surfaces under a reference illuminant, suitable allowance having been made for the state of chromatic adaptation.

Contrast: A term that is used (a) subjectively and (b) objectively:
(a) Subjective assessment of the difference in appearance of two parts of a field of view seen simultaneously or successively. Hence: brightness contrast, colour contrast.
(b) Quantity usually defined as a luminance ratio: L_o/L_b, where L_o is the object luminance and L_b the background luminance.

Contrast sensitivity: The ability of the eye to perceive the smallest difference in luminance, and thus to appreciate the slightest nuances of brightness. The contrast sensitivity is greatest when the adaptation level is between 0.01 and 1000 cd/m^2. Within this range a contrast equal to about 2% of the surrounding luminance can be observed.

Courtyard: An exterior space enclosed laterally by the walls of one or several buildings and open to the exterior at the top and sometimes laterally.

Curtain wall: A continuous translucent or transparent vertical or almost vertical surface, usually glass, with no structural function, which separates the interior from the exterior of a building. It usually does not allow ventilation.

Daylight: The combination of diffuse skylight and sunlight.

Daylight factor (DF): The illuminance received at a point indoors, expressed as a percentage of the horizontal illuminance outdoors from an unobstructed hemisphere of the same sky. Direct sunlight is excluded from both values. Consists of the sum of the sky component (SC), the externally reflected component (ERC) and the internally reflected component (IRC).

Daylighting system: A device located near or in the openings of the building envelope, whose primary function is to redirect a significant part of the incoming natural light flux (diffuse + direct) in order to improve the lighting conditions in the interior. This improvement may be to the overall daylight level, or in the distribution of daylight, or both.

Diffuse lighting: A form of lighting where approximately the same intensity of light comes from all directions.

Diffuse surface: A surface where the reflected or transmitted light flux is distributed in all directions (towards the surface of an imaginary hemisphere), although the incident beam may be direct (unidirectional).

Directional lighting: A form of lighting where the major part of light is received from a single direction.

Disability glare: Glare that impairs the vision of objects without necessarily causing discomfort.

Discomfort glare: Glare that causes discomfort without necessarily impairing the vision of objects.

Efficacy: see Luminous efficacy.

Externally reflected component (ERC): Ratio, expressed as a percentage of that part of illuminance at a point on a given plane that is received directly after reflection from external obstructions, to the illuminance on a horizontal plane due to an unobstructed hemisphere of this sky (see Daylight factor).

Fritted glass: Made by depositing and firing opaque ceramic dots (circular patches), usually between 1 and 10 mm diameter, onto the surface of glass, occupying between 30% and 70% of the surface. The dots are usually white, and thus reflect away a corresponding percentage of the radiation.

Galleria: Covered street often incorporated in shopping or exhibition areas.

Gallery: An intermediate light space attached to a building, which can be either open to the exterior or closed off by glass.

Glare: The discomfort or impairment of vision experienced when parts of the visual field are excessively bright in relation to the general surroundings.

Glare by reflection: Glare produced by specular reflections, particularly when the reflected images appear in the same or nearly the same direction as the object viewed.

Goniophotometer: A photometer for measuring the directional light distribution characteristics of sources, luminaires, media or surfaces. Spectro-goniophotometer also allows spectrally dependent measurements.

Greenhouse: An intermediate space attached to a building by one of its faces, the others being separated from the exterior by a frame supporting transparent or translucent surfaces.

High-performance glass: Glass that reduces the total energy transmission by more, in proportion, than visible radiation, thereby increasing the luminous efficacy of daylighting.

Hue: The nominal colour – the name we give to a particular sensation, e.g. red, orange, yellow, etc.

Illuminance: The luminous flux density at a surface, i.e. the luminous flux incident per unit area. Measured in lux (SI unit).

Indicatrix of diffusion or scattering indicatrix: The surface formed by the extremities of the radius vector drawn in all directions from an element of a surface, when each radius vector represents the (relative) luminous intensity or the (relative) luminance in the corresponding direction. In many cases only a meridian section of this indicatrix is required. The term indicatrix is then used to denote, instead of the surface, the curve obtained in a similar manner in a plane normal to the element concerned.

Internally reflected component (IRC): Ratio, expressed as a percentage, of that part of illuminance at a point on a given plane that is received directly after reflection from interior surfaces, to illuminance on a horizontal plane due to an unobstructed hemisphere of this sky (see Daylight factor).

Lantern: A structure, often on the highest point of a roof, with lateral openings.

Lightness constancy: The phenomenon by which the eye appears to be able to differentiate between a light surface poorly illuminated and a dark surface that receives high illumination, even though the two surfaces may have the same physical luminance.

Light-duct: A hollow linear element that conducts natural light to interior zones of a building. Its surfaces are usually finished with high-reflection materials.

Light output ratio (LOR): The ratio of the total light output of a luminaire under stated practical conditions to the 'input' from the lamp that it houses. It can be divided into downward and upward LOR.

Lightshelf: A horizontal shelf positioned (usually above eye level) to reflect daylight onto the ceiling and to shield direct glare from the sky.

Light transmittance: The fraction of visible light transmitted through the glazing and shading system if present, compared with the unglazed, unshaded aperture.

Lightwell: Toplit space penetrating one floor or more, without a wall separating the space from the surrounding spaces.

Louvres: A series of exterior or interior slats or lamellae, which may be fixed or adjustable.

Luminaire: An apparatus that controls the distribution of light given by a lamp or lamps and which includes all the components necessary for fixing and protecting the lamps and for connecting them to the supply circuit.

Luminance: The physical measure of the stimulus that produces the sensation of brightness, measured by the luminous intensity of the light emitted or reflected in a given direction from a surface element, divided by the projected area of the element in the same direction. Measured in candela / m^2 (SI unit).

Luminous efficacy: The ratio of the luminous flux emitted by a lamp, to the power consumed by the lamp. Lumens/watt (SI unit).

Luminous intensity: A quantity that describes the power of a source to emit light in a given direction. Measured in candelas; 1 candela = 1 lumen / steradian.

Lumen: The SI unit of luminous flux. 1 candela emits 4π lumens.

Lux: The SI unit of illuminance. 1 lux = 1 lumen / m^2.

Lynes ratio: If (DF$_{ave}$ front half of the room / DF$_{ave}$ back half) is greater than 3 then the back half of the room will appear unacceptably gloomy.

Monitor: A raised section of roof that includes a vertically (or near-vertically) glazed aperture for the purpose of daylight illumination. Similar to lantern but extended in one direction.

Munsell system: A system of surface colour classification using colour scales of hue, value and chroma. Now uses the terms hue, lightness and saturation.

'No-sky' line: The position on the reference plane in a room from which, because of external obstructions, there is no direct view of the sky.

Overcast sky: Sky that has 100% cloud cover (sun is not visible).

Overhang: A horizontal building projection, usually above a window, for the purpose of shading.

Reflectance: The ratio of the reflected light flux (or energy) to the incident light flux (or energy).

Rooflight: Daylight opening in the roof of a building.

Sky component (SC): Ratio, expressed as a percentage, of that part of illuminance at a point on a given plane that is received directly from a sky, to illuminance on a horizontal plane due to an unobstructed hemisphere of this sky (see Daylight factor).

Skylight: Part of solar radiation that reaches the earth's surface as a result of scattering in the atmosphere.

Skylight: Opening situated in a horizontal or tilted roof that permits the zenithal entry of daylight.

Solar altitude: Vertical angle of the sun from the horizon.

Solar control glass: Glass that reduces the light transmission and the total energy transmission.

Specular surface: If an opaque material is said to be specular, it means that it reflects like a mirror – a direct beam is reflected as a direct beam. If a transparent material is described as specular, it means that it transmits a direct beam of light without dispersing it, and it follows from this that a sharp focused view can be seen through the material.

Sunlight: That part of solar radiation that reaches the earth's surface as parallel rays, directly from the sun's disc, after selective attenuation by the atmosphere.

Sunpath diagram:
(a) Cylindrical projection: represents the azimuth, or direction of the sun from the point of interest, on a horizontal axis, and the vertical altitude of the sun on a vertical axis.
(b) Stereographic projection: represents the whole hemispherical sky dome as a circular disk with its centre corresponding to the zenith, i.e. vertically overhead, and its circumference representing the horizon. The resulting view can be likened to a 180° fish-eye photograph taken by a photographer lying on his back.

Total shading coefficient: The ratio of total transmitted radiation (including visible and invisible) passing through the window when the shading device is deployed compared with that of an unshaded, single glazed window. The qualification 'total' indicates that it includes re-radiated and convected energy from the shading elements or glass.

Translucent glass: A glass with the property of transmitting light diffusely, and through which vision varies from almost clear to almost obscure.

Transmittance: The ratio of the transmitted flux to the incident flux.

Uniformity ratio: The ratio of the minimum illuminance to the average illuminance of an area.

Uniform Sky: A hypothetical sky where the illuminance is uniform for all angles of altitude and azimuth. Often used to approximate to a low latitude sky.

Utilisation factor: The proportion of the luminous light flux emitted by the lamps that reaches the working plane; a property of both the luminaire and the room.

U-value: The thermal transmittance of a material. Units: $W / m^2 K$.

Value (Munsell): A description of the brightness of a colour on a scale from dark to light. This may vary within one colour e.g. dark red–light red, or between colours e.g. yellow (light) – violet (dark). Modern term is 'lightness'.

Visual display unit (VDU): A self-luminous screen on which information is displayed. When the VDU is part of a computer system, it is sometimes called a VDT (visual display terminal).

Veiling reflections: Specular reflections that appear on the object viewed and partially or wholly obscure the details by reducing contrast.

Visual acuity: The capacity for seeing distinctly objects very close together. It is measured as the reciprocal of the value, in minutes of arc, of the angular separation of two neighbouring objects (points or lines) which the eye can just see as separate (1 minute of arc = 1/60 degree). The ability to resolve a separation, one minute of arc wide, between two signs (acuity = 1) is often considered as 'normal' acuity. However, under adequate lighting conditions the acuity of a person with good vision should well reach double this value.

Visual comfort probability (VCP): The rating of a lighting system expressed as a percentage of the people who, when viewing from a specified location and in a specified direction, will be expected to find it comfortable in terms of discomfort glare.

Visual environment: The total space that can be seen from a particular location by moving one's head and eyes.

Visual field: The area or extent of physical space visible to an eye in a given position.

Visual perception: The quantitative assessment of impressions transmitted from the retina to the brain in terms of information about a physical world displayed before the eye.

Visual performance: Performance of the visual system as measured for instance by the speed and accuracy with which a visual task is performed.

Working plane: The plane on which the task is located. For design purposes a reference workplane is usually defined over the entire area of a room between 0.7 and 0.9 m above the floor. Lighting standards prescribe specific illuminance levels on this reference work plane according to the type of tasks that need to be performed.

Zenith: The direction normal to the horizontal plane. The zenith corresponds to an altitude of 90°

Index